Anatomy & Physiology

made

Incredibly Easy!™

Springhouse Corporation
Springhouse, Pennsylvania

Staff

Publisher
Judith A. Schilling McCann, RN, MSN

Design Director
John Hubbard

Editorial Director
Michael Shaw

Clinical Manager
Joan M. Robinson, RN, MSN, CCRN

Senior Associate Editor
Brenna H. Mayer

Editors
Cheryl Duksta, Ty Eggenberger, Kevin Haworth, Stephen Page, Kirk Robinson, Frank Thakuria

Clinical Editors
Collette Bishop Hendler, RN, CCRN (clinical project manager); Joanne Bartelmo, RN, MSN, CCRN; Gwynn Sinkinson, RN,C, MSN, CRNP; Beverly Ann Tscheschlog, RN

Copy Editors
Jaime Stockslager (supervisor), Virginia Baskerville, Mary T. Durkin, Amy Furman, Kimberly A.J. Johnson, Scotti Kent, Judith Orioli, Pamela Wingrod

Designers
Arlene Putterman (associate design director), Mary Ludwicki (art director), Joseph John Clark, Lynn Foulk

Illustrators
Michael Adams, Phillip Ashley, John Carlance, Jean Gardener, Tom Herbert, Bob Jackson, John Lymerman, Cynthia Mason, John Murphy, Judy Newhouse, Bob Neumann, Bot Roda, Betty Winnberg

Electronic Production Services
Diane Paluba (manager), Joyce Rossi Biletz

Manufacturing
Deborah Meiris (director), Patricia K. Dorshaw (manager), Otto Mezei (book production manager)

Project Coordinator
Liz Schaeffer

Editorial and Design Assistants
Tom Hasenmayer, Beverly Lane, Beth Janae Orr

Indexer
Ellen Brennan

Printed in the United States of America.

IEAP- D N O S A J J
03 02 01   10 9 8 7 6 5 4 3 2

Library of Congress Cataloging-in-Publication Data

Anatomy & physiology made incredibly easy.
    p.; cm.
    Includes index.
    1. Human physiology—Outlines, syllabi, etc. 2. Human anatomy—Outlines, syllabi, etc. I. Title: Anatomy and physiology made incredibly easy. II. Springhouse Corporation.
    [DNLM: 1. Anatomy 2. Physiology. QS 4 A5385 2000]
QP41.A53 2000
612—dc21
00-041973
ISBN 1-58255-043-3 (alk. paper)   CIP

Contents

Contributors and consultants | iv
Foreword | v

1   The human body   1

2   Chemical organization   27

3   Integumentary system   39

4   Musculoskeletal system   49

5   Nervous system   69

6   Endocrine system   101

7   Cardiovascular system   115

8   Hematologic system   131

9   Immune system   145

10   Respiratory system   161

11   Gastrointestinal system   181

12   Nutrition and metabolism   221

13   Urinary system   241

14   Fluids, electrolytes, acids, and bases   255

15   Reproductive system   275

16   Reproduction and lactation   295

Index   317

Contributors and consultants

Lisa L. Dutton, PT, MS
Physical Therapy Program Director
University of Findlay (Ohio)

Sari F. Edelstein, RD, PhD, LD, CDE
Dietitian Consultant
Private Practice
Miami

Mary Ellen Kelly, RN, BSN
Director of Staff Development and Infection Control
John L. Montgomery Care Center
Freehold, N.J.

Pamela Mullen Kovach, RN, BSN
Independent Clinical Consultant
Perkiomenville, Pa.

Anita Lockhart, RN,C, MSN
Clinical Nurse Specialist
Presbyterian Medical Center
Philadelphia

Anne Marie Palatnik, RN, MSN, CS
Clinical Nurse Specialist
Our Lady of Lourdes Medical Center
Camden, N.J.

Robert W. Russell, MS, RRT, CPFT
Regional Vice President
Respiratory Health Services, Inc.
Conshohocken, Pa.

Dianne Weyer, RN, MS, CFNP
Assistant Professor
School of Nursing
Georgia State University
Atlanta

Jacqueline H. Zaremba, RN, BSN, CCRN
RN Clinical Analyst
The Medical Center at Princeton (N.J.)

Foreword

For anyone involved in health care, a thorough knowledge of the human body is essential. Unfortunately, learning about anatomy and physiology usually requires hauling heavy textbooks home and squinting at endless diagrams of nerves and neurons, arteries and enzymes. Pouring through weighty anatomy references such as these leaves you with a solid understanding of only two things — sore muscles and tired eyes.

That is why the clinical experts at Springhouse created *Anatomy & Physiology Made Incredibly Easy*. This remarkable book will help you exercise your brain without exhausting your body (or your patience). You might even find yourself having fun.

The first chapter provides a quick overview of the human body, including cell structure and the composition of body tissues. Chapter 2 covers essential chemical processes and the molecules necessary for life. Chapters 3 through 13 discuss individual body systems. You'll find thorough discussions of each body system, complete with clear explanations and a format designed to promote learning and comprehension. Each chapter also contains vivid diagrams, charts, and illustrations — the visual tools necessary to grasp human anatomy.

But comprehending the human body isn't just about understanding distinct body parts — you also need to know how those systems work together. *Anatomy & Physiology Made Incredibly Easy* presents special chapters on nutrition and metabolism; fluid, electrolyte, and acid-base balance; and reproduction and lactation, showing how different aspects of physiology join to produce these vital processes.

Along the way, you'll also find features that enhance your learning while keeping the book lively and entertaining. *Memory joggers* provide hints for remembering key points. A *Quick quiz* at the end of each chapter helps you assess your progress. Our cast of *Incredibly Easy* characters point out essential information while providing a tongue-in-cheek approach to learning. In addition, special logos throughout alert you to important points:

Zoom in! provides a closer look at anatomic structures you need to know.

Now I get it! transforms complex elements of physiology into easy-to-digest explanations.

Body shop breaks down complex anatomy of organs and tissues.

As our understanding of the human body grows more complex, a book such as this one serves as an invaluable resource, whether you're looking to brush up your skills or studying anatomy for the first time. No more searching through complicated textbooks for quick answers at the bedside, and no more struggling to stay awake through a chapter. *Anatomy & Physiology Made Incredibly Easy* will provide a relaxed, fun guide to the human body. Finally, here is a book that will lift your knowledge — without weighing you down.

Barry S. Eckert, PhD
Associate Provost
University of Buffalo

Here's a chance to learn anatomy and physiology the *Incredibly Easy* way. No heavy lifting required!

1

The human body

Just the facts

In this chapter, you'll learn:

♦ how to use anatomic terms for direction, reference planes, body cavities, and body regions to help describe the locations of various body structures

♦ the structure of cells and how cells reproduce and generate energy

♦ the four basic tissue types and their characteristics.

Anatomic terms

Anatomic terms describe directions within the body as well as the body's planes, cavities, and regions.

Directional terms

When navigating the body, directional terms help you determine the exact location of a structure.

Couples at odds

Generally, directional terms can be grouped in pairs of opposites:

• *Superior* and *inferior* mean above and below, respectively. For example, the shoulder is superior to the elbow, and the hand is inferior to the wrist.

• *Anterior* and *posterior* mean toward the front of the body and toward the back. *Ventral* is sometimes used in-

> Locating body structures starts with directional terms, reference planes, cavities, and regions.

stead of anterior, whereas *dorsal* is sometimes used instead of posterior.

• *Medial* and *lateral* mean toward the body's midline and away from it.

• *Proximal* and *distal* mean closest and farthest, respectively, to the point of origin (or to the trunk).

• *Superficial* and *deep* mean toward or at the body surface and farthest from it.

Memory jogger

To help remember proximal and distal, keep in mind that when something is in "proximity," or nearby, it's proximal. When something is "distant," or far away, it's distal.

Reference planes

Reference planes are imaginary lines used to section the body and its organs. These lines run longitudinally, horizontally, and on an angle. (See *Body reference planes.*)

Body shop

Body reference planes

Body reference planes are used to indicate the locations of body structures. Shown here are the median sagittal, frontal, and transverse planes. An oblique plane—a slanted plane that lies between a horizontal plane and a vertical plane—isn't shown.

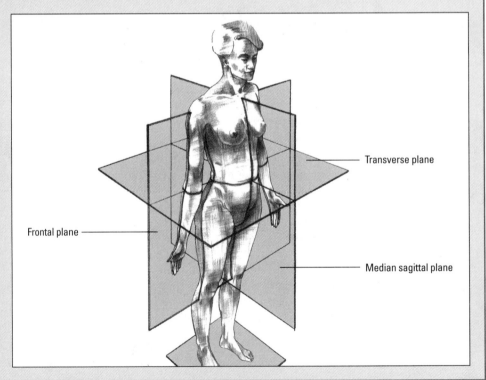

Transverse plane

Frontal plane

Median sagittal plane

Imagine this

The four major body reference planes are:

 median sagittal

 frontal

 transverse

 oblique.

Body cavities

Body cavities are spaces within the body that contain the internal organs. The *dorsal* and *ventral cavities* are the two major closed cavities — cavities without direct openings to the outside of the body. (See *Locating body cavities.*)

Body shop

Locating body cavities

The dorsal cavity, in the posterior region of the body, is divided into the cranial and vertebral cavities. The ventral cavity, in the anterior region, is divided into the thoracic and abdominopelvic cavities.

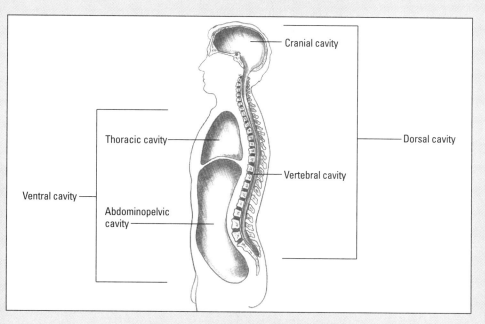

Dorsal cavity

The dorsal cavity is located in the posterior region of the body.

The think tank and backbone of the operation

The dorsal cavity is further subdivided into two cavities:
• The *cranial cavity* (also called the *calvaria*) encases the brain.
• The *vertebral cavity* (also called the *spinal cavity* or *vertebral canal*), formed by the vertebrae, encloses the spinal cord.

Ventral cavity

The ventral cavity is found in the anterior region of the trunk. This cavity is subdivided into the *thoracic cavity* and the *abdominopelvic cavity*.

Open this chest and you'll find a heart and lungs

The thoracic cavity is located superior to the abdomino-pelvic cavity and is surrounded by the ribs and chest muscles. It's subdivided into the *pleural cavities* and the *mediastinum:*
• The two *pleural cavities* each contain a lung.
• The *mediastinum* contains the heart, large vessels of the heart, trachea, esophagus, thymus, lymph nodes, and other blood vessels and nerves.

The bread basket and below

The abdominopelvic cavity is subdivided into two regions, the *abdominal cavity* and the *pelvic cavity:*
• The *abdominal cavity* contains the stomach, intestines, spleen, liver, and other organs.
• The *pelvic cavity*, inferior to the abdominal cavity, contains the bladder, some of the reproductive organs, and the rectum.

Other cavities

The body also contains an *oral cavity* (the mouth), a *nasal cavity* (located in the nose), *orbital cavities* (which house the eyes), *middle ear cavities* (which contain the small bones of the middle ear), and the *synovial cavities* (enclosed within the capsules surrounding freely moveable joints).

Hey, out there, I'm Mr. Lungs. I'm located in the pleural cavities.

Body regions

Body regions are designations for specific body areas that have a special nerve or vascular supply or those that perform a special function.

The guts of the matter

The most widely used terms are those that designate regions in the abdomen. (See *Abdominal region exposed.*) There are six regions in the abdomen:

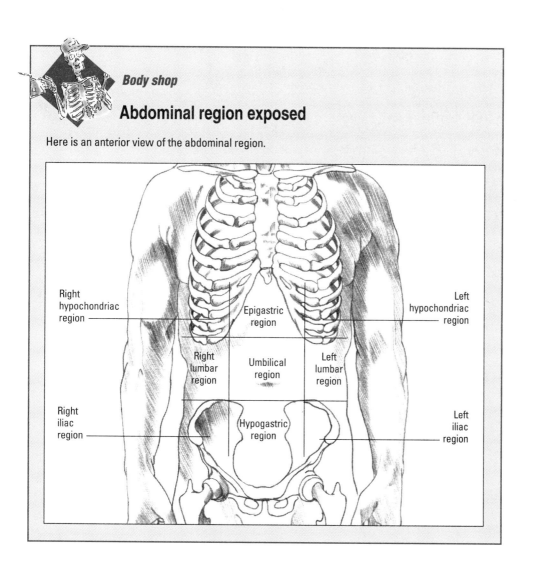

Body shop

Abdominal region exposed

Here is an anterior view of the abdominal region.

Right hypochondriac region

Epigastric region

Left hypochondriac region

Right lumbar region

Umbilical region

Left lumbar region

Right iliac region

(Hypogastric) region

Left iliac region

• The *umbilical region,* the area around the umbilicus, includes sections of the small and large intestines, inferior vena cava, and abdominal aorta.
• The *epigastric region,* superior to the umbilical region, contains most of the pancreas and portions of the stomach, liver, inferior vena cava, abdominal aorta, and duodenum.
• The *hypogastric region* (or pubic area) lies inferior to the umbilical region. Prominent structures include a portion of the sigmoid colon, urinary bladder and ureters, and portions of the small intestine.
• The right and left *iliac regions* (or inguinal regions) are situated on either side of the hypogastric region. They include portions of the small and large intestines.
• The right and left *lumbar regions* (or loin regions) are located on either side of the umbilical region. They include portions of the small and large intestines and portions of the kidneys.
• The right and left *hypochondriac regions* are located on either side of the epigastric region. They contain the diaphragm, portions of the kidneys, the right side of the liver, the spleen, and part of the pancreas.

Introducing the cell

The *cell* makes up the body's structure and serves as the basic unit of living matter. Human cells vary widely, ranging from the simple squamous epithelial cell to the highly specialized neuron.

The greatest regeneration

Generally, the simpler the cell, the greater its power to regenerate. The more specialized the cell, the weaker its regenerative power. Cells with greater regenerative power have a shorter life span than those with less regenerative power.

Cell structure

Cells are made up of three basic components:
• *protoplasm*
• *plasma membrane*
• *nucleus.* (See *Inside the cell.*)

Sometimes simpler is better. The simpler I am, the more I can regenerate!

Protoplasm

Protoplasm, a viscous, translucent, watery material, is the primary component of plant and animal cells. It contains a large percentage of water, inorganic ions (such as potassium, calcium, magnesium, and sodium), and naturally occurring organic compounds (such as proteins, lipids, and carbohydrates).

Getting charged

The inorganic ions within the protoplasm are called *electrolytes*. They regulate acid-base balance and control the amount of intracellular water. When these ions gain electrons, they acquire a positive electrical charge. When they lose electrons, they acquire a negative electrical charge.

A pair of "plasms"

Nucleoplasm is the protoplasm of the cell's nucleus. It plays a part in reproduction. *Cytoplasm* is the protoplasm of the cell body that surrounds the nucleus. It

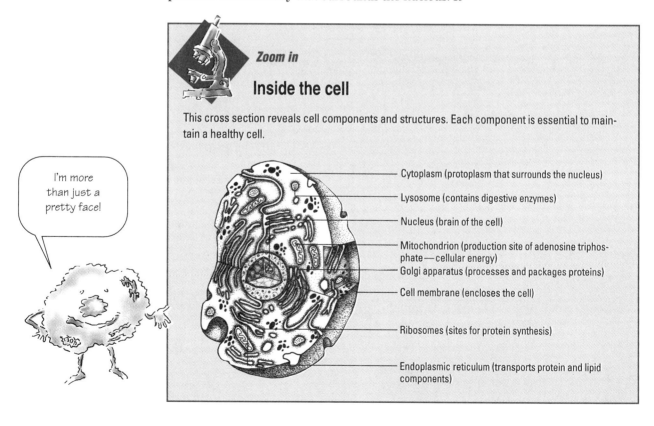

Zoom in

Inside the cell

This cross section reveals cell components and structures. Each component is essential to maintain a healthy cell.

I'm more than just a pretty face!

Cytoplasm (protoplasm that surrounds the nucleus)

Lysosome (contains digestive enzymes)

Nucleus (brain of the cell)

Mitochondrion (production site of adenosine triphosphate—cellular energy)

Golgi apparatus (processes and packages proteins)

Cell membrane (encloses the cell)

Ribosomes (sites for protein synthesis)

Endoplasmic reticulum (transports protein and lipid components)

converts raw materials to energy. It's also the site of most synthesizing activities. In the cytoplasm you'll find *cytosol*, *organelles*, and *inclusions*.

A cytosol sea

Cytosol is a viscous, semitransparent fluid that is 70% to 90% water. It contains proteins, salts, and sugars.

A lot to metabolize

Organelles are the cell's metabolic units. Each organelle performs a specific function to maintain the life of the cell:

• *Mitochondria* are threadlike structures within the cytoplasm that provide most of the body's adenosine triphosphate — the enzyme that fuels many cellular activities.
• *Ribosomes* are the sites of protein synthesis.
• The *endoplasmic reticulum* is an extensive network of membrane-enclosed tubules. *Rough endoplasmic reticulum* is covered with ribosomes and produces certain proteins. *Smooth endoplasmic reticulum* contains enzymes that synthesize lipids.
• Each *Golgi apparatus* synthesizes carbohydrate molecules. These molecules combine with the proteins produced by rough endoplasmic reticulum to form secretory products such as lipoproteins.
• *Lysosomes* are digestive bodies that break down foreign or damaged material in cells. (See *Lysosomes at work.*)
• *Peroxisomes* contain *oxidases*, enzymes capable of reducing oxygen to hydrogen peroxide and hydrogen peroxide to water.
• *Cytoskeletal elements* form a network of protein structures.
• *Centrosomes* contain *centrioles*, short cylinders that are adjacent to the nucleus and take part in cell division.

Organelles are my metabolic units.

Temps that don't do any work

Inclusions are nonfunctioning units in the cytoplasm that are commonly temporary. The pigment *melanin* in epithelial cells and the stored nutrient *glycogen* in liver cells are both examples of nonfunctioning units.

Now I get it!

Lysosomes at work

Lysosomes are the organelles responsible for digestion within the cell. Phagocytes assist in this process by engulfing foreign material and transporting it through the cell membrane. Here is how lysosomes work.

Function of lysosomes
Lysosomes are digestive bodies that break down foreign or damaged material in cells. A membrane surrounds each lysosome and separates its digestive enzymes from the rest of the cytoplasm.

Breaking it down
The lysosomal enzymes digest matter brought into the cell by *phagocytes*, special cells that surround and engulf matter outside the cell and then transport it through the cell membrane. The membrane of the lysosome fuses with the membrane of the cytoplasmic spaces surrounding the phagocytized material; this fusion allows the lysosomal enzymes to digest the engulfed material.

Plasma membrane

The *plasma membrane* (cell membrane) is the gatekeeper of the cell. It serves as the cell's external boundary, separating it from other cells and from the external environment.

Checkpoint

Nothing gets by this semipermeable membrane without authorization from the nucleus. Roughly 75Å (3 10-millionths of an inch) thick, it consists of a double layer of phospholipids with protein molecules.

Nucleus

The *nucleus* is the cell's mission control. It plays a role in cell growth, metabolism, and reproduction.

Inside the nucleus

A nucleus may contain one or more *nucleoli* — a dark-staining structure that synthesizes *ribonucleic acid* (RNA). The nucleus also contains *chromosomes*. Chromosomes control cellular activity and direct protein synthesis through ribosomes in the cytoplasm.

DNA and RNA

Protein synthesis is essential for the growth of new tissue and the repair of damaged tissue. *Deoxyribonucleic acid* (DNA) carries genetic information and provides the blueprint for protein synthesis. RNA transfers this genetic information to the ribosomes, where protein synthesis occurs.

Touching all the bases

The basic structural unit of DNA is a *nucleotide*. Nucleotides consist of a phosphate group that is linked to a five-carbon sugar, deoxyribose, and joined to a nitrogen-containing compound called a base. Four different DNA bases exist:

 adenine (A)

 guanine (G)

 thymine (T)

 cytosine (C).

We're complementary. That means we fit together perfectly.

Identifying rings

Adenine and *guanine* are double-ring compounds classified as *purines. Thymine* and *cytosine* are single-ring compounds classified as *pyrimidines.*

The chain gangs

DNA chains exist in pairs held together by weak chemical attractions between the nitrogen bases on adjacent chains. Because of the chemical shape of the bases, adenine bonds only with thymine and guanine bonds only with cytosine. Bases that can link with each other are called *complementary.*

Insider trading

RNA consists of nucleotide chains that differ slightly from the nucleotide chains found in DNA. Several types of RNA are involved in the transfer (to the ribosomes) of genetic information essential to protein synthesis. (See *Types of RNA.*)

Now I get it!

Types of RNA

There are three types of ribonucleic acid (RNA): ribosomal, messenger, and transfer. Each has its own specific function.

Ribosomal RNA
Ribosomal RNA is used to make ribosomes in the endoplasmic reticulum of the cytoplasm, where the cell produces proteins.

Messenger RNA
Messenger RNA directs the arrangement of amino acids to make proteins at the ribosomes. Its single strand of nucleotides is complementary to a segment of the deoxyribonucleic acid chain that contains instructions for protein synthesis. Its chains pass from the nucleus into the cytoplasm, attaching to ribosomes there.

Transfer RNA
Transfer RNA consists of short nucleotide chains, each of which is specific for an individual amino acid. Transfer RNA transfers the genetic code from messenger RNA for the production of a specific amino acid.

Chromosomes

Chromosomes are composed of DNA and protein. They appear as a network of *chromatin granules* in the nondividing cell. Chromosomes exist in pairs except in the *gametes* (germ cells). In the gametes, one chromosome comes from the male germ cell; the other, from the female germ cell.

22 identical twins...

Normal human cells contain 23 pairs of chromosomes. In these cells, 22 pairs are called *homologous chromosomes*. These chromosomes are identical in size, shape, and gene location and each chromosome in a pair contains genetic information that controls the same characteristics or functions.

...and a 23rd pair that determines gender

The 23rd pair contains *sex* (X and Y) *chromosomes*. In the 23rd pair, one chromosome comes from the female germ cell (ovum) and the other from the male germ cell (spermatozoon). The combination of these chromosomes determines gender: XX produces a genetic female; XY produces a genetic male.

An X-factor

In the female, the genetic activity of both X chromosomes is essential only during the first few weeks after conception. Later development requires just one functional X chromosome. The other X chromosome is inactive and appears as a dense chromatin mass called a *Barr body* (sex chromatin body). It's attached to the nuclear membrane in the cells of a normal female.

As a woman, I have two X chromosomes. One X chromosome becomes inactive and appears as a Barr body.

Genes

Genes are segments of chromosomal DNA chains, arranged in a line on the chromosomes, that determine the properties of a cell. The *gene locus* is the location of a specific gene on a chromosome. *Alleles* are alternate forms of a gene that can occupy a particular gene locus; only one allele can occupy a specific locus.

Our designer genes

Because chromosomes are paired, genes also occur in pairs on homologous chromosomes, with one allele at each of the four gene loci. If the alleles for a particular gene are the same on both chromosomes, the person is *homozygous* for that gene. If the alleles differ, the person is *heterozygous* for that gene. (See *How genes express themselves.*)

Cell reproduction

Cells are under a constant call to reproduce; it's either that or die. Cell division is how cells reproduce (or replicate) themselves; they achieve this through the process of *mitosis* or *meiosis*.

DNA does its thing

Before a cell divides, its chromosomes are duplicated. During this process, the double helix separates into two DNA chains. Each chain serves as a template for constructing a new chain. Individual DNA nucleotides are linked into new strands with bases complementary to those in the original.

Now I get it!

How genes express themselves

Genes account for inherited traits. *Gene expression* refers to a gene's effect on cell structure or function; however, the effects vary with the gene.

Dominant genes

If genes could speak, dominant genes would be loud and garrulous, dominating every conversation! Dominant genes (such as the one for dark hair) can be expressed and transmitted to the offspring even if only one parent possesses the gene.

Recessive genes

Unlike dominant genes, recessive genes prefer to hide their light under a bushel basket. A recessive gene (such as the one for blond hair) is expressed only when both parents transmit it to the offspring.

Codominant genes

Firm believers in equality, codominant genes, as in the genes that direct specific types of hemoglobin synthesis in red blood cells, allow expression of both alleles.

Sex-linked genes

Sex-linked genes are carried on sex chromosomes. Almost all appear on the X chromosome and are recessive. In the male, sex-linked genes behave like dominant genes because no second X chromosome exists.

Double double

In this way, two identical double helices are formed, each containing one of the original strands and a newly formed complementary strand. These double helices are duplicates of the original DNA chain. (See *DNA up close*, page 14.)

Mitosis

Mitosis is the equal division of material in the nucleus (*karyokinesis*) followed by division of the cell body (*cytokinesis*). It's the preferred mode of replication by all cells in the human body, except the gametes. Cell division occurs in five phases, an inactive phase called *interphase*, and four active phases:

 prophase

 metaphase

 anaphase

 telophase.

Zoom in

DNA up close

Linked deoxyribonucleic acid (DNA) chains form a spiral structure, or *double helix*.

A spiral staircase

To understand linked DNA chains, imagine a spiral staircase. The deoxyribose and phosphate groups form the railings of the staircase, and the nitrogen base pairs (adenine and thymine, guanine and cytosine) form the steps.

Cell division

Each chain serves as a template for constructing a new chain. When a cell divides, individual DNA nucleotides are linked into new strands with bases complementary to those in the originals. In this way, two identical double helices are formed, each containing one of the original strands and a newly formed complementary strand. These double helices are duplicates of the original DNA chain.

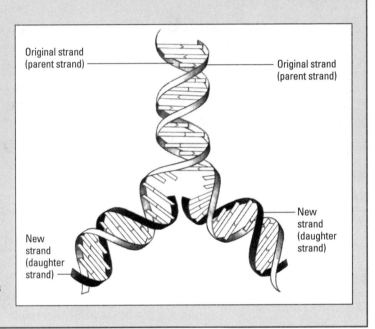

Original strand (parent strand)

Original strand (parent strand)

New strand (daughter strand)

New strand (daughter strand)

Two daughters equal 46

Mitosis results in two daughter cells (exact duplicates), each containing 23 pairs of chromosomes — or 46 individual chromosomes. This number is the *diploid number*. (See *Divide and conquer: Five stages of mitosis.*)

Meiosis

Meiosis is reserved for gametes (ova and spermatozoa). This process intermixes genetic material between homologous chromosomes, producing four daughter cells, each with the *haploid number* of chromosomes (23 or half of the 46). Meiosis has two divisions separated by a resting phase.

First division

The first division has six phases and begins with one parent cell. When the first division ends, the result is two

Now I get it!

Divide and conquer: Five stages of mitosis

Through the process of mitosis, the nuclear content of all body cells (except gametes) reproduces and divides. The result is the formation of two new daughter cells, each containing the diploid (46) number of chromosomes.

Interphase

During *interphase*, the nucleus and nuclear membrane are well defined, and the nucleolus is visible. As chromosomes replicate, each forms a double strand that remains attached at the center by a centromere.

Prophase

In *prophase*, the nucleolus disappears and the chromosomes become distinct. *Chromatids*, halves of each duplicated chromosome, remain attached by the centromere. Centrioles move to opposite sides of the cell and radiate spindle fibers.

Metaphase

Metaphase occurs when chromosomes line up randomly in the center of the cell between the spindles, along the *metaphase plate.* The centromere of each chromosome then replicates.

Anaphase

Anaphase is characterized by centromeres moving apart, pulling the separate chromatids (now called *chromosomes*) to opposite ends of the cell. The number of chromosomes at each end of the cell equals the original number.

Telophase

During *telophase,* the final stage of mitosis, a nuclear membrane forms around each end of the cell and spindle fibers disappear. The cytoplasm compresses and divides the cell in half. Each new cell contains the diploid (46) number of chromosomes.

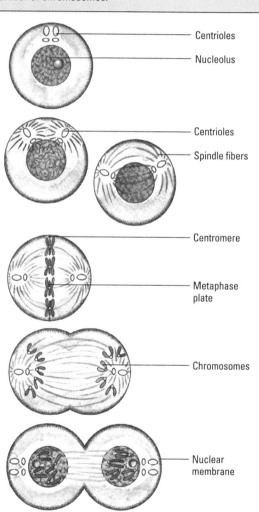

Centrioles
Nucleolus

Centrioles
Spindle fibers

Centromere

Metaphase plate

Chromosomes

Nuclear membrane

daughter cells — each containing the haploid (23) number of chromosomes.

Division 2, the sequel

The second division is a four-phase division that resembles mitosis. It starts with two new daughter cells, each containing the haploid number of chromosomes, and ends with four new haploid cells. In each cell, the two chromatids of each chromosome separate to form new daughter cells. However, because each cell entering the second division has only 23 chromosomes, each daughter cell formed has only 23 chromosomes. (See *Meiosis: Step-by-step.*)

Cellular energy generation

All cellular function depends on energy generation and transportation of substances within and among cells.

Powering cellular processes

Adenosine triphosphate (ATP) serves as the chemical fuel for cellular processes. ATP consists of a nitrogen-containing compound (adenine) joined to a five-carbon sugar (ribose), forming adenosine. Adenosine is joined to three phosphate (or triphosphate) groups. Chemical bonds between the first and second phosphate groups and between the second and third phosphate groups contain abundant energy.

The three r's

ATP needs to be converted to *adenosine diphosphate* (ADP) to produce energy. To understand this conversion, remember the three "r's" — rupture, release, and recycle.
- *Rupture:* ATP is converted to ADP when the terminal high-energy phosphate bond ruptures.
- *Release:* Because the third phosphate is liberated, energy stored in the chemical bond is released.
- *Recycle:* Mitochondrial enzymes then reconvert ADP and the liberated phosphate to ATP. To obtain the energy needed for this reattachment, mitochondria oxidize food nutrients. This makes recycled ATP available again for energy production.

Now I get it!

Meiosis: Step-by-step

Meiosis has two divisions separated by a resting phase. By the end of the first division, there are two daughter cells that each contain the haploid (23) number of chromosomes. When the second division ends, each of the two daughter cells from the first division divides, resulting in four daughter cells, each containing the haploid number of chromosomes.

First division
First division has six phases. Here is what happens during each one.

Interphase
1. Chromosomes replicate, forming a double strand attached at the center by a centromere.
2. Chromosomes appear as an indistinguishable matrix within the nucleus.
3. Centrioles appear outside the nucleus.

Prophase I
1. Nucleolus and nuclear membrane disappear.
2. Chromosomes are distinct, with chromatids attached by the centromere.
3. Homologous chromosomes move close together and intertwine; exchange of genetic information (genetic recombination) may occur.
4. Centrioles separate and spindle fibers appear.

Metaphase I
1. Pairs of synaptic chromosomes line up randomly along the metaphase plate.
2. Spindle fibers attach to each chromosome pair.

Anaphase I
1. Synaptic pairs separate.
2. Spindle fibers pull homologous, double-stranded chromosomes to opposite ends of the cell.
3. Chromatids remain attached.

Telophase I
1. Nuclear membrane forms.
2. Spindle fibers and chromosomes disappear.
3. Cytoplasm compresses and divides the cell in half.
4. Each new cell contains the haploid (23) number of chromosomes.

Interkinesis
1. Nucleus and nuclear membrane are well defined.
2. Nucleolus is prominent, and each chromosome has two chromatids that don't replicate.

Second division
Second division closely resembles mitosis and is characterized by these four phases.

Prophase II
1. Nuclear membrane disappears.
2. Spindle fibers form.
3. Double-stranded chromosomes appear as thin threads.

Metaphase II
1. Chromosomes line up along the metaphase plate.
2. Centromeres replicate.

Anaphase II
1. Chromatids separate (now a single-stranded chromosome).
2. Chromosomes move away from each other to the opposite ends of the cell.

Telophase II
1. Nuclear membrane forms.
2. Chromosomes and spindle fibers disappear.
3. Cytoplasm compresses, dividing the cell in half.
4. Four daughter cells are created, each of which contains the haploid (23) number of chromosomes.

Movement within the cells

Each cell interacts with body fluids through the interchange of substances.

Modes of transportation

Several transport methods — *diffusion, osmosis, active transport,* and *endocytosis* — move substances between cells and body fluids. In another method, *filtration,* fluids and dissolved substances are transferred across capillaries into *interstitial fluid* (fluid in the spaces between cells and tissues).

No energy is required for diffusion. So I'll just go with the flow.

Diffusion

In *diffusion,* solutes move from an area of higher concentration to one of lower concentration. Eventually an equal distribution of solutes between the two areas occurs.

Go with the flow

Diffusion is a form of passive transport — no energy is required to make it happen; it just happens. It's kind of like fish traveling downstream. They just go with the flow.

Advancing and declining rates

Several factors influence the rate of diffusion:
• *concentration gradient* — the greater the concentration gradient (the difference in particle concentration on either side of the plasma membrane), the faster diffusion takes place
• *particle size* — the smaller the particles are, the faster the rate of diffusion
• *lipid solubility* — the more lipid-soluble the particles are, the more rapidly they diffuse through the lipid layers of the cell membrane.

Osmosis

Osmosis is the passive movement of fluid across a membrane from an area of lower solute concentration (comparatively *more* fluid) into an area of higher solute concentration (comparatively *less* fluid).

Enough is enough

Osmosis stops when enough fluid has moved through the membrane to equalize the solute concentration on both sides of the membrane. (See *Osmosis: Fluid moves.*)

Now I get it!

Osmosis: Fluid moves

In osmosis, fluid moves passively from an area with more fluid (and fewer solutes) to one with less fluid (and more solutes). Remember that in osmosis *fluid* moves, not solutes.

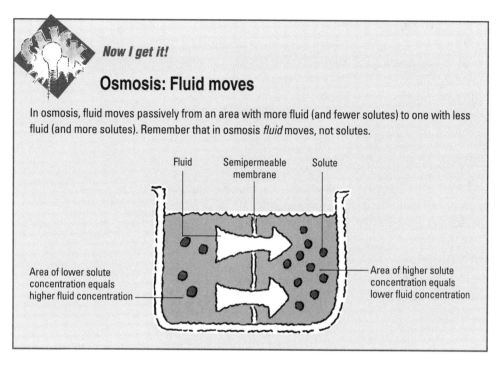

Fluid   Semipermeable   Solute
         membrane

Area of lower solute concentration equals higher fluid concentration

Area of higher solute concentration equals lower fluid concentration

Active transport

Active transport requires energy. Usually, this mechanism moves a substance across the cell membrane against the concentration gradient — from an area of lower concentration to one of higher concentration. Think of active transport as swimming upstream. When a fish swims upstream, it has to expend energy.

ATP at it again

The energy required for a solute to move against a concentration gradient comes from ATP. ATP is stored in all cells and supplies energy for solute movement in and out of cells.

It goes both ways

However, active transport also can move a substance with the concentration gradient. In this process, a carrier molecule in the cell membrane combines with the substance and transports it through the membrane, depositing it on the other side.

Active transport requires energy. As a cell, that means I need to fill up on ATP.

Endocytosis

Endocytosis is an active transport method in which, instead of passing through the cell membrane, a substance is engulfed by the cell. The cell surrounds the substance with part of the cell membrane. This part separates to form a *vacuole* (cavity) that moves to the cell's interior.

Gobbling up particles

Endocytosis involves either *phagocytosis* or *pinocytosis*. Phagocytosis refers to engulfment and ingestion of particles too large to pass through the cell membrane. Pinocytosis occurs only to engulf dissolved substances or small particles suspended in fluid.

Filtration

Fluid and dissolved substances also may move across a cell membrane by *filtration*.

Pressure is the point

In filtration, pressure (provided by capillary blood) is applied to a solution on one side of the cell membrane. The pressure forces fluid and dissolved particles through the membrane. The rate of filtration (how quickly substances pass through the membrane) depends on the amount of pressure. Filtration promotes the transfer of fluids and dissolved materials from the blood across the capillaries into the interstitial fluid.

Introducing human tissue

Tissues are groups of cells that perform the same general function. The human body contains four basic types: *epithelial, connective, muscle,* and *nervous tissue*.

Epithelial tissue

Epithelial tissue (epithelium) is a continuous cellular sheet that covers the body's surface, lines body cavities, and forms certain glands. Imagining a mummy wrapped in strips of cloth will give you some idea of how epithe-

A crowd of us cells working together on the same function is called tissue.

lial tissue covers the human body. (See *Distinguishing types of epithelial tissue*, page 22.)

Patrolling the borders

Some columnar epithelial cells in the lining of the intestines have vertical striations, forming a *striated border*. In the tubules of the kidneys, borders of columnar epithelial cells have tiny, brushlike structures (microvilli) called a *brush border*.

This hair isn't just for looks

Two common types of cells that form epithelial tissue are *stereociliated* and *ciliated epithelial cells*. The former line the epididymis and have long, piriform (pear-shaped) tufts. *Ciliated epithelial cells* possess *cilia*, fine hairlike protuberances. Cilia are larger than microvilli and move fluid and particles through the cavity of an organ.

Lining of the heart

Epithelial tissue with a single layer of squamous cells attached to a basement membrane is called *endothelium*. Such tissue lines the heart, lymphatic vessels, and blood vessels.

Glandular epithelium

Organs that produce secretions consist of a special type of epithelium called *glandular epithelium*.

The secret is in how it secretes

Glands are classified as exocrine or endocrine according to how they secrete their products:
• *Endocrine glands* release their secretions into the blood or lymph. For instance, the medulla of the adrenal gland secretes epinephrine and norepinephrine into the bloodstream.
• *Exocrine glands* discharge their secretions onto external or internal surfaces. An example is sweat, which sweat glands secrete onto the surface of the skin.

Mixing it up

Mixed glands contain both endocrine and exocrine cells. The pancreas is a mixed gland. It contains alpha and beta cells (in the islets of Langerhans). These endocrine cells produce glucagon and insulin, respectively.

Zoom in

Distinguishing types of epithelial tissue

Epithelial tissue (epithelium) is classified by the number of cell layers and the shape of surface cells. Some types of epithelium are *desquamated,* or shed, and regenerate continuously by transformation of cells from deeper layers.

Identified by number of cell layers

Classified by number of cell layers, epithelium may be *simple* (one-layered), *stratified* (multilayered), or *pseudostratified* (one-layered but appearing to be multilayered).

Classified by shape

If classified by shape, epithelium may be *squamous* (containing flat surface cells), *columnar* (containing tall, cylindrical surface cells), or *cuboidal* (containing cube-shaped surface cells).

The top left illustration shows how the basement membrane of simple squamous epithelium joins the epithelium to underlying connective tissues. The remaining illustrations show the five other types of epithelial tissue.

Simple squamous epithelium

Single layer of flattened cells with disc-shaped nuclei

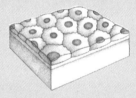

Simple columnar epithelium

Single layer of tall cells with oval nuclei

Stratified squamous epithelium

Basal cells that are cuboidal or columnar

Simple cuboidal epithelium

Single layer of cubelike cells

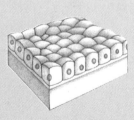

Stratified columnar epithelium

Superficial cells that are elongated and columnar

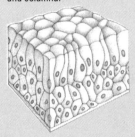

Pseudostratified columnar epithelium

Cells of different height with nuclei at different levels

Connective tissue

Connective tissue — a category that includes bone, cartilage, and adipose (fatty) tissue — binds together and supports body structures. Connective tissue is classified as *loose* or *dense*.

Cut loose

Loose (areolar) connective tissue has large spaces that separate the fibers and cells. It contains a lot of intercellular fluid.

Dense tissue issues

Dense connective tissue provides structural support and has greater fiber concentration. Dense tissue is further subdivided into dense regular and dense irregular connective tissue:

• *Dense regular* connective tissue consists of tightly packed fibers arranged in a consistent pattern. It includes tendons, ligaments, and *aponeuroses* (flat fibrous sheets that attach muscles to bones or other tissues).

• *Dense irregular* connective tissue has tightly packed fibers arranged in an inconsistent pattern. It's found in the dermis, submucosa of the GI tract, fibrous capsules, and fasciae.

Often called fat

Commonly called fat, *adipose tissue* is a specialized type of loose connective tissue where a single lipid (fat) droplet occupies most of each cell. Widely distributed subcutaneously, it acts as insulation to conserve body heat, as a cushion for internal organs, and as a storage depot for excess food and reserve supplies of energy.

Muscle tissue

Muscle tissue consists of muscle cells with a generous blood supply. Muscle cells measure up to several centimeters long and have an elongated shape that enhances their *contractility* (ability to contract).

The tissues at issue

There are three basic types of muscle tissue:
• *Striated muscle tissue* gets its name from its striped, or striated, appearance; it contracts voluntarily.
• *Cardiac muscle tissue* is sometimes classified as striated because it too is composed of striated tissue. However, it differs from other striated muscle tissue in two ways: its fibers are separate cellular units that don't contain many nuclei, and it contracts involuntarily.
• *Smooth-muscle tissue* consists of long, spindle-shaped cells and lacks the striped pattern of striated tissue. Its activity is stimulated by the autonomic nervous system and isn't under voluntary control.

Wall paper

Smooth-muscle tissue lines the walls of many internal organs and other structures, including:
• the respiratory passages from the trachea to the alveolar ducts
• the urinary and genital ducts
• the arteries and veins
• the larger lymphatic trunks
• the arrectores pilorum
• the iris and ciliary body of the eye.

Nervous tissue

The main function of *nervous tissue* is communication. Its primary properties are *irritability* (the capacity to react to various physical and chemical agents) and *conductivity* (the ability to transmit the resulting reaction from one point to another).

Irritable!?
I'll show you irritable!

Nervous tissue specialists

Neurons are highly specialized cells that generate and conduct nerve impulses. A typical neuron consists of a cell body with cytoplasmic extensions — numerous *dendrites* on one pole and a single *axon* on the opposite pole. These extensions allow the neuron to conduct impulses over long distances.

Protecting neurons

Neuroglia form the support structure of nervous tissue, insulating and protecting neurons. They're found only in the central nervous system.

Quick quiz

1. The reference plane that divides the body lengthwise into right and left regions is the:
    A. frontal plane.
    B. sagittal plane.
    C. transverse plane.

Answer: B. Imaginary lines called reference planes are used to section the body. The sagittal plane runs lengthwise and divides the body into right and left regions.

2. The organelle that plays the biggest role in cellular function is the:
    A. nucleus.
    B. Golgi apparatus.
    C. ribosomes.

Answer: A. Serving as the cell's control center, the nucleus plays a role in cell growth, metabolism, and reproduction.

3. The four basic types of tissue that the human body contains are:
    A. epithelial, connective, muscle, and nervous tissue.
    B. bone, cartilage, glands, and adipose tissue.
    C. loose, dense connective, dense regular, and dense irregular tissue.

Answer: A. Tissues are groups of cells with the same general function. The human body contains four basic types of tissue: epithelial, connective, muscle, and nervous tissue.

4. In DNA, bases that can link with each other are called complementary. Which of the following are complementary bases?
    A. Adenine and guanine
    B. Cytosine and thymine
    C. Adenine and thymine

Answer: C. Because of the chemical shape of the bases, adenine bonds only with thymine, and guanine bonds only with cytosine.

5. According to genetic theory, if a child has blond hair this must mean:

     A.   both parents transmit the gene for blond hair.

     B.   one parent transmits the gene for blond hair.

     C.   one grandparent has blond hair.

Answer: A. Because the trait for blond hair is a recessive gene, it's only expressed when both parents transmit it to the offspring.

6. Meiosis ends when:

     A.   two new daughter cells form, each with the haploid number of chromosomes.

     B.   one daughter cell forms and is an exact copy of the original.

     C.   four new daughter cells form, each with the haploid number of chromosomes.

Answer: C. Meiosis comes to completion with the end of telophase II. The result is four daughter cells, each of which contains the haploid (23) number of chromosomes.

Scoring

☆☆☆   If you answered all six questions correctly, fantastic! You're well on your way to a fantastic voyage through the human body.

☆☆   If you answered four or five questions correctly, all right! You're in for some smooth sailing and pretty soon you'll know the body inside and out.

☆   If you answered fewer than four questions correctly, buck up, camper. With fifteen more quick quizzes, you'll conquer this body of knowledge in no time.

2

Chemical organization

Just the facts

In this chapter, you'll learn:

♦ the chemical composition of the body

♦ the structure of an atom

♦ the difference between inorganic and organic compounds.

The body's chemistry

The human body is composed of chemicals; in fact, all of its activities are chemical in nature. Understanding chemistry is essential to understanding the human body and its functions.

It all comes down to chemistry

The chemical level is the simplest and most important level of structural organization. Without the proper chemicals in the proper amounts, body cells — and eventually the body itself — would die.

Principles of chemistry

Every cell contains thousands of different chemicals that constantly interact with one another. Differences in chemical composition differentiate types of body tissue. What is more, the blueprints of heredity (deoxyribonucleic acid [DNA] and ribonucleic acid [RNA]) are encoded in chemical form.

Without chemicals in the proper amounts, I would die.

Now I get it!

Understanding elements and compounds

It can be confusing to understand the difference between elements and compounds. The best way to remember the difference is to understand how atoms combine to form each.

Get at them atoms
A single atom constitutes an element. Thus, an atom of hydrogen is the element Hydrogen. Now, here is the confusing part: An element can also be composed of more than one atom — a molecule.

Molecules: Two (or more) of the same
A molecule is a combination of two or more atoms. If these atoms are the same — that is, all the same element (such as all hydrogen atoms) — they're considered a *molecule* of that element (a molecule of hydrogen).

Compounds: Where the different come together
If the atoms combined are different — that is, different elements (such as a carbon atom and an oxygen atom) — the molecule formed is a *compound* (such as the compound CO, or carbon monoxide).

> Yep. I'm an atom — the smallest unit of matter that can take part in a chemical reaction.

> We're atoms that are joined together to make a molecule. We're the same, so we're a molecule of an element.

> I know we just met, but I think we'll make a beautiful compound together.

What's the matter?

Matter is anything that has mass and occupies space. It may be a solid, liquid, or gas.

Energetic types

Energy is the capacity to do work — to put mass into motion. It may be *potential energy* (stored energy) or *kinetic energy* (the energy of motion). Types of energy include chemical, electrical, and radiant.

Chemical composition

An *element* is matter that can't be broken down into simpler substances by normal chemical reactions. All forms of matter are composed of chemical elements. Each of the chemical elements in the periodic table has a chemical symbol. For example, Ca is the chemical symbol for calcium. (See *Understanding elements and compounds.*)

It's elementary

Carbon, hydrogen, nitrogen, and oxygen account for 96% of the body's total weight. Calcium and phosphorus account for another 2.5%. (See *What is a body made of?* page 30.)

Atomic structure

An *atom* is the smallest unit of matter that can take part in a chemical reaction. Atoms of a single type constitute an element.

Subatomic particles

Each atom has a dense central core called a *nucleus*, plus one or more surrounding energy layers called *electron shells*. Atoms consist of three basic subatomic particles: *protons*, *neutrons*, and *electrons*.

Relative weights

A proton weighs nearly the same as a neutron, and a proton and a neutron each weigh 1,836 times as much as an electron.

Protons

Protons (p+) are closely packed particles in the atom's nucleus that have a positive charge. Each element has a distinct number of protons.

Positive thinking

An element's number of protons determines its *atomic number* and positive charge. For example, all carbon atoms — and *only* carbon atoms — have 6 protons; therefore, the atomic number of carbon is 6 (6p+).

Body shop

What is a body made of?

This chart shows the chemical elements of the human body in descending order from most to least plentiful.

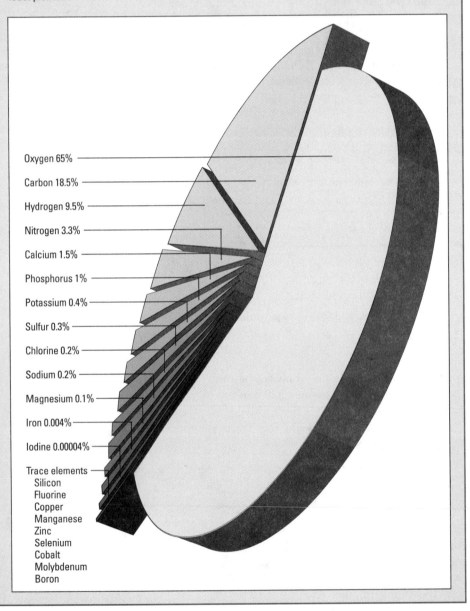

Oxygen 65%

Carbon 18.5%

Hydrogen 9.5%

Nitrogen 3.3%

Calcium 1.5%

Phosphorus 1%

Potassium 0.4%

Sulfur 0.3%

Chlorine 0.2%

Sodium 0.2%

Magnesium 0.1%

Iron 0.004%

Iodine 0.00004%

Trace elements
  Silicon
  Fluorine
  Copper
  Manganese
  Zinc
  Selenium
  Cobalt
  Molybdenum
  Boron

With all that oxygen, it's a wonder I don't just float away.

Neutrons

Neutrons (n) are uncharged, or neutral, particles in the atom's nucleus.

Mass numbers

An atom's *atomic mass number* is distinct from its atomic number. The atomic mass number is the sum of the number of protons and neutrons in the nucleus of an atom. You can also think of the atomic mass number as the sum of the masses of protons and neutrons. For example, helium, with two protons and two neutrons, has an atomic mass number of four.

Isolating the isotopes

Not all the atoms of an element necessarily have the same number of neutrons. *Isotopes* are forms of an atom with a different number of neutrons (and a different atomic weight) than most atoms of the element.

A weighty matter

Understanding isotopes is a key to another important concept, *atomic weight*. An atom's atomic weight is the average of the relative weights (atomic mass numbers) of all the element's isotopes. (Recall that isotopes are different atomic forms of the same element that vary in the number of neutrons they contain.)

Electrons

Electrons (e-) are negatively charged particles that orbit the nucleus in electron shells. They play a key role in chemical bonds and reactions.

Staying neutral

The number of electrons in an atom equals the number of protons in its nucleus. The electrons' negative charges cancel out the protons' positive charges, making atoms electrically neutral.

Shell games

Electrons circle the nucleus in *shells* or concentric circles. Each electron shell can hold a maximum number of electrons and represents a specific energy level. The innermost shell can accommodate two electrons at most, whereas the outermost shells can hold many more.

An atom with single (unpaired) electrons orbiting in its outermost electron shell can be chemically active — that is, able to take part in chemical reactions. An atom with an outer shell that contains only pairs of electrons is chemically inactive, or stable.

The value of valence

An atom's *valence* (its ability to combine with other atoms) equals the number of unpaired electrons in its outer shell. For example, sodium (Na+) has a plus-one valence because its outer shell contains an unpaired electron.

Chemical bonds

A *chemical bond* is a force of attraction that binds a molecule's atoms together. Formation of a chemical bond usually requires energy. Breakup of a chemical bond usually releases energy.

The name's bond...

Several types of chemical bonds exist:
• A *hydrogen bond* occurs when two atoms associate with a hydrogen atom. Oxygen and nitrogen, for instance, commonly form hydrogen bonds.
• An *ionic* (electrovalent) *bond* occurs when valence electrons transfer from one atom to another.
• A *covalent bond* forms when atoms share pairs of valence electrons. (See *Picturing ionic and covalent bonds.*)

Chemical reactions

A *chemical reaction* involves unpaired electrons in the outer shells of atoms. In this reaction, one of two events occurs:

☝ Unpaired electrons from the outer shell of one atom transfer to the outer shell of another atom.

✌ One atom shares its unpaired electrons with another atom.

How will they react?

Energy, particle concentration, speed, and orientation determine whether a chemical reaction will occur. The four basic types of chemical reactions are *synthesis, decom-*

Now I get it!

Picturing ionic and covalent bonds

A chemical bond is a force of attraction that binds the atoms of a molecule together. Let's examine ionic and covalent bonds.

Ionic bonds

In an ionic bond, an electron is transferred from one atom to another. By forces of attraction, an electron is transferred from a sodium (Na) atom to a chlorine (Cl) atom. The result is a molecule of sodium chloride (NaCl).

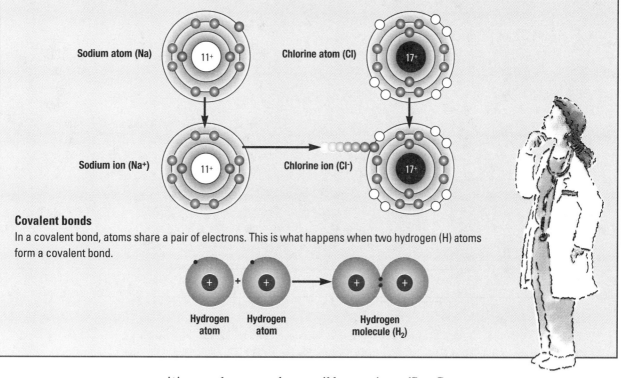

Covalent bonds

In a covalent bond, atoms share a pair of electrons. This is what happens when two hydrogen (H) atoms form a covalent bond.

position, exchange, and reversible reactions. (See Comparing chemical reactions, page 34.)

Inorganic and organic compounds

Although most biomolecules (molecules produced by living cells) form *organic compounds*, or compounds containing carbon, some form *inorganic compounds*, or compounds without carbon.

Now I get it!

Comparing chemical reactions

When chemical reactions occur, they involve unpaired electrons in the outer shells of atoms. Here are the four basic types of chemical reactions.

Synthesis reaction (anabolism)

A synthesis reaction combines two or more substances (reactants) to form a new, more complex substance (product). This results in a chemical bond.

$$A + B \rightarrow AB$$

Decomposition reaction (catabolism)

In a decomposition reaction, a substance decomposes, or breaks down, into two or more simpler substances, leading to the breakdown of a chemical bond.

$$AB \rightarrow A + B$$

Exchange reaction

An exchange reaction is a combination of a decomposition and a synthesis reaction. This reaction occurs when two complex substances decompose into simpler substances. The simple substances then join (through synthesis) with different simple substances to form new complex substances.

$$AB + CD \rightarrow A + B + C + D \rightarrow AD + BC$$

Reversible reaction

In a reversible reaction, the product reverts to its original reactants, and vice versa. Reversible reactions may require special conditions, such as heat or light.

$$A + B \leftrightarrow AB$$

Inorganic compounds

Inorganic compounds are usually small and include water and *electrolytes* — inorganic acids, bases, and salts.

The body's reservoir

Water is the body's most abundant substance. It performs a host of vital functions, including:
- easily forming polar covalent bonds (which permits the transport of solvents)
- acting as a lubricant in mucus and other bodily fluids
- entering into chemical reactions, such as nutrient breakdown during digestion
- enabling the body to maintain a relatively constant temperature (by both absorbing and releasing heat slowly).

Recognizing ionizing

Acids, bases, and salts are *electrolytes* — compounds whose molecules consist of positively charged ions, *cations*, and negatively charged ions, *anions*, that *ionize* (separate into ions) in solution.

- *Acids* ionize into hydrogen ions (H+) and anions. In other words, acids separate into a positively charged hydrogen ion and a negatively charged anion.
- *Bases*, in contrast, ionize into hydroxide ions and cations. Bases separate into negatively charged hydroxide ions and positively charged cations.
- *Salts* form when acids react with bases. In water, salts ionize into cations and anions, but not hydrogen or hydroxide ions.

I need water to stay at a constant temperature.

A balancing act

Body fluids must attain acid-base balance to maintain *homeostasis* (the dynamic equilibrium of the body). A solution's acidity is determined by the number of hydrogen ions it contains. The more hydrogen ions present, the more acidic the solution. Conversely, the more hydroxide ions a solution contains, the more basic, or *alkaline*, it is.

Organic compounds

Most biomolecules form *organic compounds* — compounds that contain carbon or carbon-hydrogen bonds. *Carbohydrates*, *lipids*, *proteins*, and *nucleic acids* are all examples of organic compounds.

Carbohydrates

In the body, *carbohydrates* are sugars, starches, glycogen, and cellulose.

The energy company

The main functions of carbohydrates are to release energy and store energy. There are three types of carbohydrates:

Monosaccharides, such as ribose and deoxyribose, are sugars with three to seven carbon atoms.

✌ *Disaccharides*, such as lactose and maltose, contain two monosaccharides.

☝ *Polysaccharides*, such as glycogen, are large carbohydrates with many monosaccharides.

Lipids

Lipids are water-insoluble biomolecules. The major lipids are *triglycerides, phospholipids, steroids, lipoproteins,* and *eicosanoids.*

I'm a carbohydrate. Count on me to release energy.

To insulate and protect

Triglycerides are the most abundant lipid in both food and the body. These lipids are neutral fats that insulate and protect. They also serve as the body's most concentrated energy source. Triglycerides contain three molecules of a fatty acid chemically joined to one molecule of glycerol.

Bars on the cell

Phospholipids are the major structural components of cell membranes and consist of one molecule of glycerol, two molecules of a fatty acid, and a phosphate group.

No fat in cholesterol?

Steroids are simple lipids with no fatty acids in their molecules. They fall into four main categories, each of which performs different functions:
• *Bile salts* emulsify fats during digestion and aid absorption of the fat-soluble vitamins (vitamins A, D, E, and K).
• *Male and female sex hormones* are responsible for sexual characteristics and reproduction.
• *Cholesterol,* a part of animal cell membranes, is needed to form all other steroids.
• *Vitamin D* helps regulate the body's calcium concentration.

Porters and other hardworking lipids

Lipoproteins help transport lipids to various parts of the body. *Eicosanoids* include *prostaglandins,* which, among other functions, modify hormone responses, promote the inflammatory response, and open the airways; and *leukotrienes,* which also play a part in allergic and inflammatory responses.

Memory jogger

To recall the distinction between the three types of carbohydrates, remember the prefixes.

Mono- means one. *Di-* means two; therefore, disaccharides contain two monosaccharides. *Poly-* means many, so you can expect that polysaccharides contain many monosaccharides.

Proteins

Proteins are the most abundant organic compound in the body. They're composed of building blocks called *amino acids*. Amino acids are linked together by *peptide bonds* — chemical bonds that join the carboxyl group of one amino acid to the amino group of another.

Building up the blocks

Many amino acids linked together form a *polypeptide*. One or more polypeptides form a protein. The sequence of amino acids in a protein's polypeptide chain dictates its shape. A protein's shape determines which of its many functions it performs:

- providing structure and protection
- promoting muscle contraction
- transporting various substances
- regulating processes
- serving as enzymes (the largest group of proteins, which act as catalysts for crucial chemical reactions).

I'm a protein. I'm built from blocks of amino acids that form polypeptides.

Nucleic acids

The nucleic acids DNA and RNA are composed of nitrogenous bases, sugars, and phosphate groups. The primary hereditary molecule, DNA contains two long chains of deoxyribonucleotides, which coil into a double-helix shape.

Holding it together

Deoxyribose and phosphate units alternate in the "backbone" of the chains. Holding the two chains together are base pairs of adenine-thymine and guanine-cytosine.

RNA and its special function

Unlike DNA, RNA has a single-chain structure. It contains ribose instead of deoxyribose and replaces the base thymine with uracil. RNA transmits genetic information from the cell nucleus to the cytoplasm. In the cytoplasm, it guides protein synthesis from amino acids.

Quick quiz

1. The three most plentiful chemical elements in the human body are:
- A. phosphorous, hydrogen, and oxygen.
- B. carbon, oxygen, and silicon.
- C. oxygen, carbon, and hydrogen.

Answer: C. There are 22 chemical elements in the human body. Oxygen (65%), carbon (18.5%), and hydrogen (9.5%) are the three most plentiful.

2. Protons are closely packed particles in the atom's nucleus that have:
- A. a positive charge.
- B. a negative charge.
- C. a neutral charge.

Answer: A. Protons are positively charged particles in the atom's nucleus.

3. An example of an organic compound is:
- A. water.
- B. an electrolyte.
- C. a protein.

Answer: C. Organic compounds include carbohydrates, lipids, proteins, and nucleic acids.

Scoring

☆☆☆ If you answered all three questions correctly, congratulations. You and this chapter have achieved homeostasis.

☆☆ If you answered two questions correctly, excellent. You and this chapter go together as neatly as amino acids forming a polypeptide chain.

☆ If you answered fewer than two questions correctly, no worries. Fourteen more quick quizzes to go!

Integumentary system

Just the facts

In this chapter, you'll learn:

♦ the basic functions of the skin

♦ the skin layers and their components

♦ the appendages (hair, nails, and glands) of the integumentary system.

Introducing the integumentary system

The integumentary system is the largest body system and includes the skin, or *integument*, and its appendages (the hair, nails, and certain glands).

Not just another pretty face

The integumentary system performs many vital functions, including:

 protection of inner body structures

 sensory perception

 regulation of body temperature

 excretion of some body fluids.

Protection

The skin maintains the integrity of the body surface by migration and shedding. It can repair surface wounds by intensifying normal cell replacement mechanisms. The skin's top layer also protects the body against noxious chemicals and invasion from bacteria and microorganisms.

Langerhans' cells to the rescue

Langerhans' cells are specialized cells in this skin layer. They enhance the body's immune response by helping lympho-cytes to process antigens entering the skin.

The skin's own sun block

Melanocytes, another type of skin cell, protect the skin by producing the brown pigment *melanin* to help filter ultraviolet light (irradiation). Exposure to ultraviolet light can stimulate melanin production.

Sensory perception

Sensory nerve fibers originate in the nerve roots along the spine and supply specific areas of the skin known as *dermatomes*.

Just sensational

These nerve fibers transmit various sensations, such as temperature, touch, pressure, pain, and itching, from the skin to the central nervous system. Autonomic nerve fibers carry impulses to smooth muscle in the walls of the skin's blood vessels, to the muscles around the hair roots, and to the sweat glands.

Body temperature

Abundant nerves, blood vessels, and eccrine glands with-in the skin's deeper layer aid thermoregulation, or control of body temperature.

Warming up...

When the skin is exposed to cold or the internal body temperature falls, blood vessels constrict, decreasing blood flow and thereby conserving body heat.

Now I get it!

The skin's role in thermoregulation

Abundant nerves, blood vessels, and eccrine glands within the skin's deeper layer aid *thermoregulation* (control of body temperature). The first flow chart shows how the body conserves body heat. The second flow chart shows how the body reduces body heat. Here is how the skin does its job.

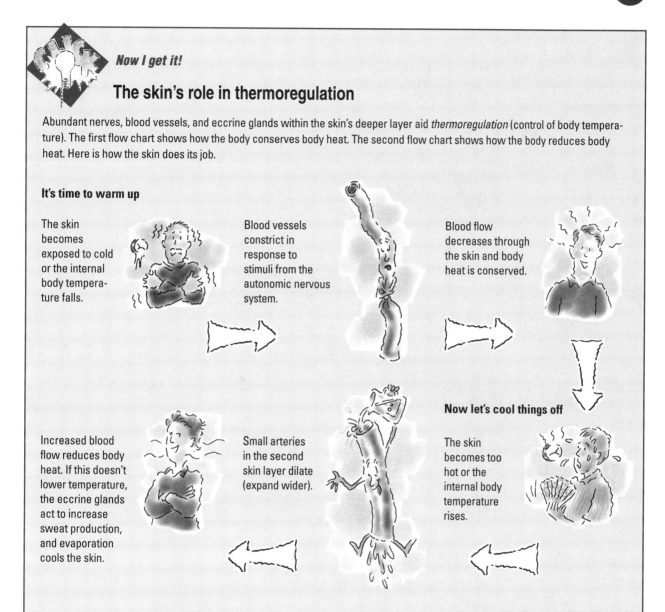

It's time to warm up

The skin becomes exposed to cold or the internal body temperature falls.

Blood vessels constrict in response to stimuli from the autonomic nervous system.

Blood flow decreases through the skin and body heat is conserved.

Now let's cool things off

The skin becomes too hot or the internal body temperature rises.

Small arteries in the second skin layer dilate (expand wider).

Increased blood flow reduces body heat. If this doesn't lower temperature, the eccrine glands act to increase sweat production, and evaporation cools the skin.

...and cooling down

If the skin becomes too hot or the internal body temperature rises, small arteries within the skin dilate, increasing the blood flow, which in turn reduces body heat. (See *The skin's role in thermoregulation.*)

Excretion

The skin is also an excretory organ. The sweat glands excrete sweat, which contains water, electrolytes, urea, and lactic acid.

Water works

While it eliminates body wastes through its more than two million pores, the skin also prevents body fluids from escaping. Here, the skin is again protecting the body by preventing dehydration caused by loss of internal body fluids—as well as maintaining these levels by regulating the content and volume of sweat. It also keeps unwanted fluids in the environment from entering the body.

Skin layers

Two distinct layers of skin, the *epidermis* and *dermis*, lie above a third layer of *subcutaneous tissue*—sometimes called the *hypodermis*. (See *A view of the skin.*)

Epidermis

The *epidermis* is the outermost layer and varies in thickness from less than 0.1 mm on the eyelids to more than 1 mm on the palms and soles.

The ins and outs (and in betweens)

The epidermis is composed of avascular, stratified, squamous (scaly or platelike) epithelial tissue and is divided into five distinct layers. Each layer is named for its structure or function:
• The *stratum corneum* is the outermost layer and consists of tightly arranged layers of cellular membranes and keratin.
• The *stratum lucidum*, or clear layer, blocks water penetration or loss. It may be missing in some thin skin.
• The *stratum granulosum*, or granular layer, is responsible for keratin formation and like the stratum lucidum may be missing in some thin skin.
• The *stratum spinosum*, or spiny layer, also helps with keratin formation and is rich in ribonucleic acid.

Zoom in

A view of the skin

Major components of the skin include the epidermis, dermis, and epidermal appendages.

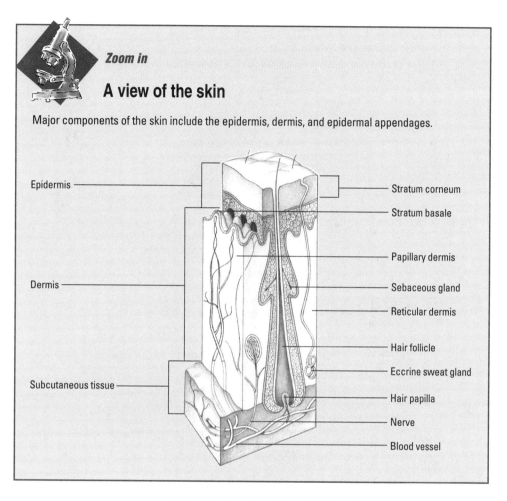

Epidermis

Dermis

Subcutaneous tissue

Stratum corneum
Stratum basale
Papillary dermis
Sebaceous gland
Reticular dermis
Hair follicle
Eccrine sweat gland
Hair papilla
Nerve
Blood vessel

• The *stratum basale*, or the basal layer, is the innermost layer and produces new cells to replace the superficial keratinized cells that are continuously shed or worn away.

Dermis

The *dermis*, also called the *corium*, is the skin's second layer. It's an elastic system that contains and supports blood vessels, lymphatic vessels, nerves, and the epidermal appendages.

What's in the matrix?

Most of the dermis is made up of extracellular material called *matrix*. Matrix contains:

Memory jogger

You can keep straight which skin layer is which by remembering that the prefix *epi-* means "upon." Therefore, the *epidermis* is upon, or on top of, the *dermis*.

- *collagen,* a protein that gives strength to the dermis
- *elastin,* which makes the skin pliable
- *reticular fibers* that bind the collagen and elastin fibers together.

The myth of fingerprints

The dermis itself has two layers:
- The *papillary dermis* has fingerlike projections, *papillae,* that connect the dermis to the epidermis. It contains characteristic ridges that on the fingers are known as fingerprints. These ridges also help the fingers and toes in gripping surfaces.
- The *reticular dermis* covers a layer of subcutaneous tissue, insulating the body to conserve heat. It also provides energy and serves as a mechanical shock absorber.

Epidermal appendages

Numerous epidermal appendages occur throughout the skin. They include the hair, nails, sebaceous glands, and sweat glands.

Hair

The *hairs* are long, slender shafts composed of keratin. At the expanded lower end of each hair is a bulb or root. On its undersurface, the root is indented by a *hair papilla,* a cluster of connective tissue and blood vessels.

It will literally make your hair stand on end

Each hair lies within an epithelium-lined sheath called a *hair follicle.* A bundle of smooth-muscle fibers, *arrector pili,* extends through the dermis to attach to the base of the follicle. When these muscles contract, the hair stands on end. Hair follicles also have a rich blood and nerve supply.

Nails

The *nails* are situated over the distal surface of the end of each finger and toe. Nails are specialized types of keratin.

When arrector pili muscles contract, hair stands on end.

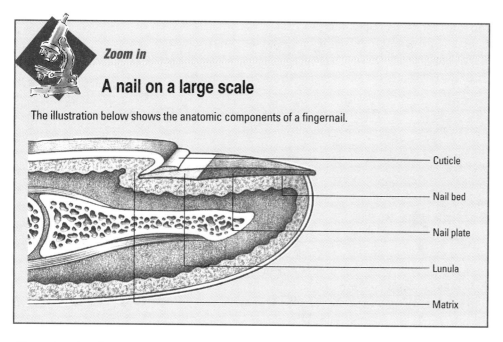

Zoom in

A nail on a large scale

The illustration below shows the anatomic components of a fingernail.

Cuticle

Nail bed

Nail plate

Lunula

Matrix

On a bed of nails

The *nail plate*, surrounded on three sides by the nail folds, or *cuticles*, lies on the nail bed. The nail plate is formed by the nail matrix, which extends proximally for about ¼″ (0.6 cm) beneath the nail fold.

Landing on the lunula

The distal portion of the matrix shows through the nail as a pale crescent-moon-shaped area. This is called the *lunula*. The translucent nail plate distal to the lunula exposes the nail bed. The vascular bed imparts the characteristic pink appearance under the nails. (See *A nail on a large scale.*)

Sebaceous glands

Sebaceous glands occur on all parts of the skin except the palms and soles. They are most prominent on the scalp, face, upper torso, and genitalia.

A (small) miracle oil

The sebaceous glands produce *sebum*, a mixture of keratin, fat, and cellulose debris. Combined with sweat, sebum forms a moist, oily, acidic film that is mildly antibacterial and antifungal and that protects the skin surface. Sebum exits through the hair follicle opening to reach the skin surface.

Eccrine glands secrete fluid in response to stress.

Sweat glands

There are two types of sweat glands:

- eccrine glands
- apocrine glands.

Eccrine glands

The *eccrine glands* are widely distributed throughout the body and produce an odorless, watery fluid with a sodium concentration equal to that of plasma. A duct from the coiled secretory portion passes through the dermis and epidermis, opening onto the skin surface.

Stressed out

Eccrine glands in the palms and soles secrete fluid mainly in response to emotional stress. For example, your eccrine glands might secrete fluid while you're taking a test. The remaining three million eccrine glands respond primarily to thermal stress, effectively regulating temperature.

Apocrine glands

The *apocrine glands* are located chiefly in the axillary (underarm) and anogenital (groin) areas. They have a coiled secretory portion that lies deeper in the dermis than that of the eccrine glands. A duct connects an apocrine gland to the upper portion of the hair follicle.

Oh, no, b.o.

Apocrine glands begin to function at puberty. However, they have no known biological function. As bacteria decompose the fluids produced by these glands, body odor occurs.

Quick quiz

1. The main functions of the skin include:
   A. support, nourishment, and sensation.
   B. protection, sensory perception, and temperature regulation.
   C. fluid transport, sensory perception, and aging regulation.

Answer: B. The skin's main functions involve protection from injury, noxious chemicals, and bacterial invasion; sensory perception of touch, temperature, and pain; and regulation of body heat.

2. The outermost layer of the skin is the:
   A. epidermis.
   B. dermis.
   C. hypodermis.

Answer: A. The outermost layer of the skin, composed of avascular, stratified, squamous epithelial tissue, is the epidermis.

3. The appendages of the skin include all of the following except:
   A. nails.
   B. hair.
   C. nerves.

Answer: C. The appendages of the skin are the nails, hair, sebaceous glands, eccrine glands, and apocrine glands.

4. Sebum is a mixture of:
   A. cellulose debris, fat, and keratin.
   B. collagen and elastin.
   C. watery fluid and sodium.

Answer: A. Sebum is produced by the sebaceous glands and is a mixture of keratin, fat, and cellulose debris.

5. The sweat glands that are widely distributed throughout the body are:
   A. apocrine.
   B. eccrine.
   C. adipose.

Answer: B. Eccrine glands are widely distributed throughout the body and produce an odorless, watery fluid.

Scoring

☆☆☆   If you answered all five questions correctly, amazing! You've got the integumentary system covered.

☆☆   If you answered three or four questions correctly, awesome. You're scratching beneath the surface of this body system.

☆   If you answered fewer than three questions correctly, no sweat. There are still thirteen more quick quizzes to go!

Musculoskeletal system

Just the facts

In this chapter, you'll learn:

♦ types of muscle tissue and their function

♦ major muscles and bones of the body

♦ role of tendons, ligaments, cartilage, joints, and bursae.

Musculoskeletal system

The musculoskeletal system consists of muscles, tendons, ligaments, bones, cartilage, joints, and bursae. These structures give the human body its shape and ability to move.

How the body moves

The various parts of the musculoskeletal system work with the nervous system to produce voluntary movements. Muscles contract when stimulated by impulses from the nervous system.

Using the force

During contraction, the muscle shortens, pulling on the bones to which it's attached. Force is applied to the tendon; then one bone is pulled toward, moved away from, or rotated around a second bone, depending on the type of muscle that has contracted. Most movement involves groups of muscles rather than one muscle.

Structures of the musculoskeletal system work together to provide support and produce movement.

Muscles

There are three major types of muscle in the human body. They're classified by the tissue they contain:

cardiac (heart) muscle, which is made up of a specialized type of striated tissue

visceral (involuntary) muscle, which contains smooth-muscle tissue

skeletal (voluntary and reflex) muscle, which consists of striated tissue.

The attached type

This chapter discusses only skeletal muscle — the type attached to bone. The human body has about 600 skeletal muscles. (See *Viewing the major skeletal muscles.*)

Muscle functions

Skeletal muscles move body parts or the body as a whole. They're responsible for both voluntary and reflex movements. Skeletal muscles also maintain posture and generate body heat.

Muscle structure

Skeletal muscle is composed of large, long cell groups called *muscle fibers.* Each fiber has many nuclei and a series of increasingly smaller internal fibrous structures. (See *Muscle structure up close*, page 52.)

Oh my, myofibrils

The structures of a muscle fiber, working from the cell's exterior to its interior, are:
• *endomysium,* a sheath of fibrous connective tissue that surrounds the exterior of the fiber
• *sarcolemma,* which is the plasma membrane of the cell that lies beneath the endomysium and just above the cells' nuclei
• *sarcoplasm,* the muscle cell's cytoplasm, which is contained within the sarcolemma

Body shop

Viewing the major skeletal muscles

This illustration shows anterior and posterior views of some of the major muscles.

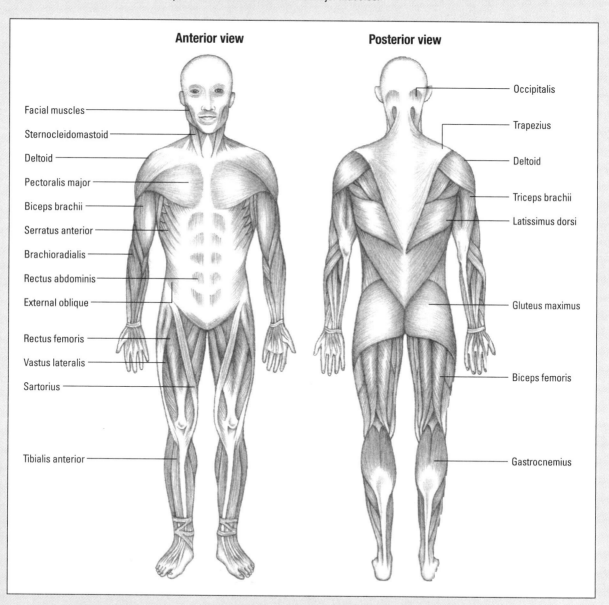

Anterior view

- Facial muscles
- Sternocleidomastoid
- Deltoid
- Pectoralis major
- Biceps brachii
- Serratus anterior
- Brachioradialis
- Rectus abdominis
- External oblique
- Rectus femoris
- Vastus lateralis
- Sartorius
- Tibialis anterior

Posterior view

- Occipitalis
- Trapezius
- Deltoid
- Triceps brachii
- Latissimus dorsi
- Gluteus maximus
- Biceps femoris
- Gastrocnemius

Zoom in

Muscle structure up close

Skeletal muscle contains cell groups called muscle fibers. The illustration below shows the muscle and its fibers.

In a bind

The *perimysium* — a sheath of connective tissue — binds muscle fibers together into a fasciculus. The *epimysium* binds the fasciculi (bundles of muscle fibers) together; beyond the muscle, it becomes a tendon.

Muscle fibers are surrounded

A sarcolemma (plasma membrane) surrounds each muscle fiber. Tiny myofibrils within the muscle fibers contain even finer fibers called myosin (thick filaments) and actin (thin filaments).

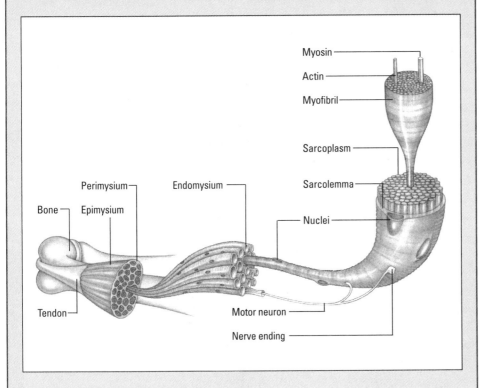

• *myofibrils*, which are tiny, threadlike structures that run the fiber's length and make up the bulk of the fiber
• *myosin* (thick filaments) and *actin* (thin filaments), which are still finer fibers within the myofibrils — about 1,500 myosin and about 3,000 actin.

Sarcomeres end to end

Myosin and actin are contained within compartments called *sarcomeres*. Sarcomeres are the functional units of skeletal muscle. During muscle contraction, myosin and actin slide over each other, reducing sarcomere length.

Zebra stripes

The sarcomere compartments of all the myofibrils in a single fiber are aligned. So, when a muscle fiber is viewed microscopically, transverse (at right angles to the long axis) stripes, called *striations*, appear along the length of the fiber.

A bundle of bundles

A fibrous sheath of connective tissue, called the *perimysium*, binds muscle fibers into a bundle, or *fasciculus*. A stronger sheath, the *epimysium*, binds all of the fasciculi together to form the entire muscle. Extending beyond the muscle, the epimysium becomes a tendon.

Muscle attachment

Most skeletal muscles are attached to bones, either directly or indirectly.

The direct approach

In a direct attachment, the epimysium of the muscle fuses to the *periosteum*, the fibrous membrane covering the bone.

Being indirect

In an indirect attachment (most common), the epimysium extends past the muscle as a tendon, or *aponeurosis*, and attaches to the bone.

Moments of contraction

During contraction, one of the bones to which the muscle is attached stays relatively stationary while the other is pulled in toward the stationary one.

Origin and insertion

The point where the muscle attaches to the stationary or less movable bone is called the *origin;* the point where it attaches to the more movable bone is called the *inser-*

> When I contract my muscles to lift this cup, one bone stays stationary. The other bone is pulled toward the stationary one.

tion. The origin usually lies on the proximal end of the bone. The insertion site is on the distal end.

Factors such as genetic constitution and exercise cause muscle strength and size to differ among individuals.

Muscle growth

Muscle develops when existing muscle fibers hypertrophy. Muscle strength and size differ among individuals because of such factors as exercise, nutrition, gender, and genetic constitution. Changes in nutrition or exercise cause muscle strength and size to vary in an individual.

Muscle movements

Skeletal muscle can permit several types of movement. A muscle's functional name comes from the type of movement it permits. For example, a flexor muscle permits bending *(flexion)*; an adductor muscle permits moving away from a body axis *(adduction)*; and a circumductor muscle allows a circular movement *(circumduction)*. (See *Basics of body movement.*)

Muscles of the axial skeleton

The muscles of the axial skeleton are essential for respiration. The axial skeleton includes the following muscles:
• muscles of the face, tongue, and neck
• muscles of mastication
• muscles of the vertebral column situated along the spine.

Muscles of the appendicular skeleton

The appendicular skeleton includes the muscles of the:
• shoulder
• abdominopelvic cavity
• upper and lower extremities.
    Muscles of the upper extremities are classified according to the bones they move. Those that move the arm are further categorized into those with an origin on the axial skeleton and those with an origin on the scapula.

Body shop

Basics of body movement

Basic muscle movement is best demonstrated in the diarthrodial joints, which allow 13 angular and circular movements:
- The shoulder demonstrates circumduction.
- The elbow demonstrates flexion and extension.
- The hip demonstrates internal and external rotation.
- The arm demonstrates abduction and adduction.
- The hand demonstrates supination and pronation.
- The jaw demonstrates retraction and protraction.
- The foot demonstrates eversion and inversion.

Retraction and protraction
Moving backward and forward

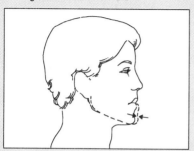

Extension
Straightening, increasing the joint angle

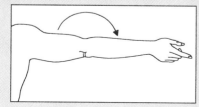

Flexion
Bending, decreasing the joint angle

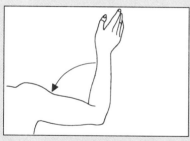

Abduction
Moving away from the midline

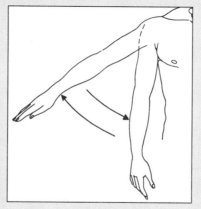

Adduction
Moving toward the midline

Circumduction
Moving in a circular manner

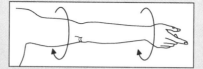

Pronation
Turning downward

Supination
Turning upward

Internal rotation
Turning toward the midline

External rotation
Turning away from the midline

Eversion
Turning outward

Inversion
Turning inward

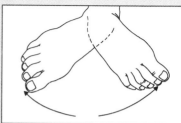

Tendons and ligaments

Tendons are bands of fibrous connective tissue that attach muscles to the periosteum, the fibrous covering of the bone. Tendons enable bones to move when skeletal muscles contract.

Ligaments are dense, strong, flexible bands of fibrous connective tissue that bind bones to other bones.

Ligaments, which connect the joint ends of bones, either limit or promote movement and also provide stability.

Bones

The human skeleton contains 206 bones: 80 form the axial skeleton — called axial because it lies along the central line, or axis, of the body — and 126 form the appendicular skeleton — relating to the limbs, or appendages, of the body. (See *Viewing the major bones.*)

Access the axis

Bones of the axial skeleton include:
- the facial and cranial bones
- the hyoid bone
- vertebrae
- ribs and sternum.

Appendages to the axis

Bones of the appendicular skeleton include:
- clavicle
- scapula
- humerus, radius, ulna, carpals, metacarpals, and phalanges
- pelvic bone
- femur, patella, fibula, tibia, tarsals, metatarsals, and phalanges.

Appendages are a good thing!

You think you're humerus, don't you?

Bone classification

Bones are typically classified by shape.
Thus, bones may be classified as:
- long (such as the humerus, radius, femur, and tibia)
(see *Viewing a long bone*, page 58)

Body shop

Viewing the major bones

The illustrations below show major bones and bone groups.

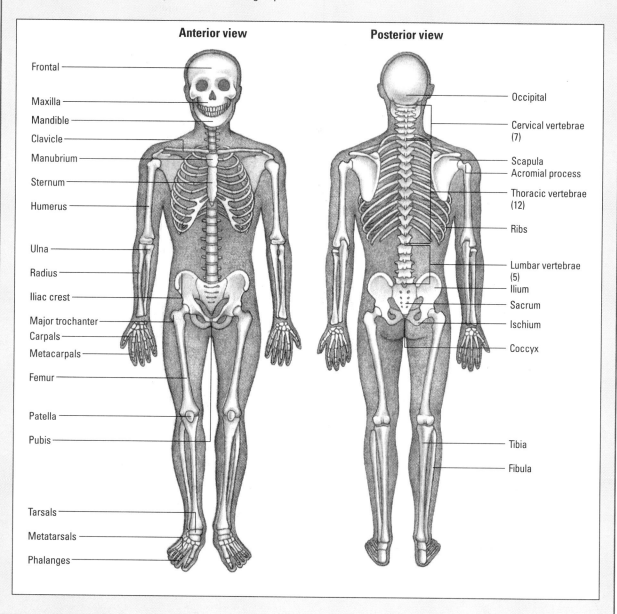

Anterior view

Frontal
Maxilla
Mandible
Clavicle
Manubrium
Sternum
Humerus
Ulna
Radius
Iliac crest
Major trochanter
Carpals
Metacarpals
Femur
Patella
Pubis
Tarsals
Metatarsals
Phalanges

Posterior view

Occipital
Cervical vertebrae (7)
Scapula
Acromial process
Thoracic vertebrae (12)
Ribs
Lumbar vertebrae (5)
Ilium
Sacrum
Ischium
Coccyx
Tibia
Fibula

Zoom in

Viewing a long bone

The main parts of a long bone, as shown below, are the *diaphysis* (shaft) and the *epiphyses* (ends). Periosteum surrounds the diaphysis; endosteum lines the medullary cavity. At the epiphyseal line, cartilage separates the epiphyses from the diaphysis.

Two types of bone tissue

Each bone consists of an outer layer of dense, smooth compact bone, which contains haversian canals, and an inner layer of spongy cancellous bone, which lacks these canals.

Cancellous bone

Cancellous bone consists of tiny spikes, called *trabeculae,* that interlace to form a latticework. Red marrow fills the spaces between the trabeculae of some bones. Cancellous bone fills the central regions of the epiphyses and the inner portions of short, flat, and irregular bones.

Compact bone

Compact bone is found in the diaphyses of long bones and the outer layers of short, flat, and irregular bones. Compact bone consists of layers of calcified matrix containing spaces occupied by osteocytes (bone cells). *Lamellae* (bone layers) are arranged around central canals (haversian canals). Small cavities called *lacunae,* which lie between the lamellae, contain osteocytes. *Canaliculi* (tiny canals) connect the lacunae, forming the structural units of the bone. Canaliculi also provide nutrients to bone tissue.

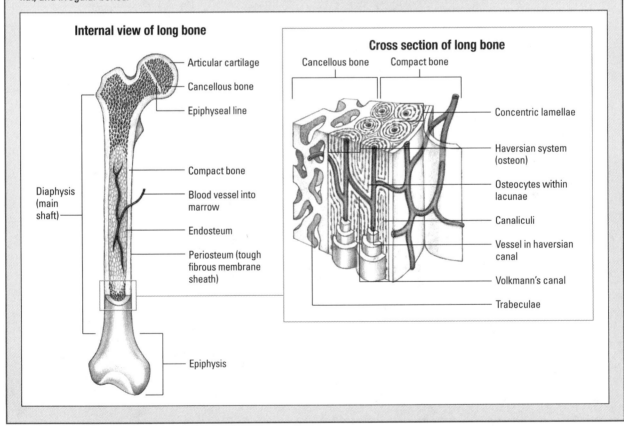

Internal view of long bone

- Articular cartilage
- Cancellous bone
- Epiphyseal line
- Compact bone
- Blood vessel into marrow
- Endosteum
- Periosteum (tough fibrous membrane sheath)

Diaphysis (main shaft)

Epiphysis

Cross section of long bone

Cancellous bone   Compact bone

- Concentric lamellae
- Haversian system (osteon)
- Osteocytes within lacunae
- Canaliculi
- Vessel in haversian canal
- Volkmann's canal
- Trabeculae

- short (such as the carpals and tarsals)
- flat (such as the scapula, ribs, and skull)
- irregular (such as the vertebrae and mandible)
- sesamoid, meaning "resembling a sesame seed" (such as the patella).

Bone functions

Bones perform various anatomic (mechanical) and physiologic functions. They:
- protect internal tissues and organs — for example, the 33 vertebrae surround and protect the spinal cord
- stabilize and support the body
- provide a surface for muscle, ligament, and tendon attachment
- move through "lever" action when contracted
- produce red blood cells in the bone marrow (a process called *hematopoiesis*)
- store mineral salts — for example, approximately 99% of the body's calcium.

I support and stabilize the body.

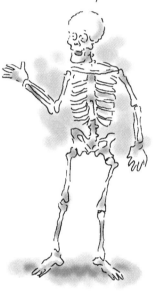

Blood supply

Blood reaches bones through three paths:

 haversian canals, minute channels that lie parallel to the axis of the bone, that are passages for arterioles

Volkmann's canals, which contain vessels that connect one Haversian canal to another and to the outer bone

vessels in the bone ends and within the marrow.

Bone formation

At 3 months in utero, the fetal skeleton is composed of cartilage. By about 6 months, fetal cartilage has been transformed into bony skeleton. (See *Bone growth and remodeling*, pages 60 and 61.)

(Text continues on page 62.)

Zoom in

Bone growth and remodeling

The ossification of cartilage into bone, or *osteogenesis,* begins at about the 9th week of fetal development. The diaphyses of long bones are formed by birth, and the epiphyses begin to ossify about that time. Below are the stages of bone growth and remodeling of the epiphyses of a long bone.

Creation of an ossification center

At about the 9th month, an ossification center develops in the epiphysis. Some cartilage cells enlarge and stimulate ossification of surrounding cells. The enlarged cells die, leaving small cavities. New cartilage continues to develop.

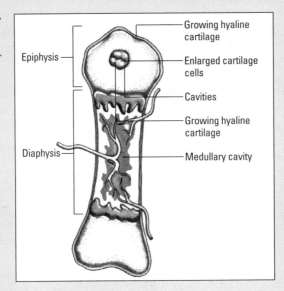

Osteoblasts form bone

Osteoblasts begin to form bone on the remaining cartilage, creating the trabeculae network of cancellous bone. Cartilage continues to form on the outer surfaces of the epiphysis and along the upper surface of the epiphyseal plate.

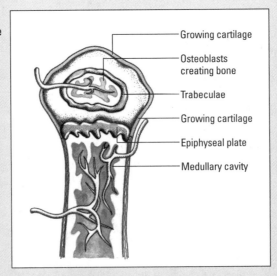

Bone growth and remodeling *(continued)*

Bone length grows

Cartilage is replaced by compact bone near the outer surfaces of the epiphysis. Only cartilage cells on the upper surface of the diaphyseal plate continue to multiply rapidly, pushing the epiphysis away from the diaphysis. This new cartilage ossifies, creating trabeculae on the medullary side of the epiphyseal plate.

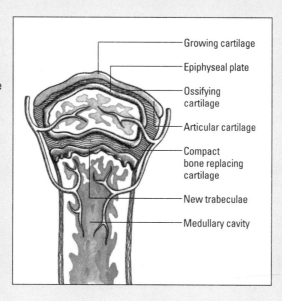

Growing cartilage

Epiphyseal plate

Ossifying cartilage

Articular cartilage

Compact bone replacing cartilage

New trabeculae

Medullary cavity

Remodeling

Osteoclasts produce enzymes and acids that reduce trabeculae created by the epiphyseal plate, thus enlarging the medullary cavity. In the epiphysis, osteoclasts reduce bone, making its calcium available for new osteoblasts that give the epiphysis its adult shape and proportion. In young adults, the epiphyseal plate completely ossifies (closes) and becomes the epiphyseal line; longitudinal growth of bone then ceases.

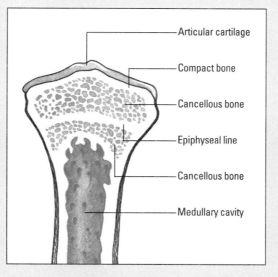

Articular cartilage

Compact bone

Cancellous bone

Epiphyseal line

Cancellous bone

Medullary cavity

Ossification is hard work

After birth, some bones — most notably the carpals and tarsals — *ossify* (harden). The change results from *endochondral ossification*, a process by which *osteoblasts* (bone-forming cells) produce *osteoid* (a collagenous material that ossifies).

Bone remodeling

Two types of osteocytes, osteoblasts and osteoclasts, are responsible for *remodeling* — the continuous process whereby bone is created and destroyed.

Blast them bones

Osteoblasts deposit new bone, and *osteoclasts* increase long-bone diameter. Osteoclasts promote longitudinal bone growth by reabsorbing the previously deposited bone. This growth continues until the *epiphyseal plates* ossify during late adolescence. The epiphyseal plates are cartilage that separate the *diaphysis*, or shaft of a bone, from the *epiphysis*, or end of a bone.

Cartilage

Cartilage is a dense connective tissue that consists of fibers embedded in a strong, gel-like substance. Unlike rigid bone, cartilage has the flexibility of firm plastic.

Cartilage supports and shapes various structures, such as the auditory canal and the intervertebral disks. It also cushions and absorbs shock, preventing direct transmission to the bone. Cartilage has no blood supply or innervation.

Bone density decreases after age 30 in women and after age 45 in men.

There are three types of cartilage:

 hyaline

 fibrous

 elastic.

Common connector

Hyaline cartilage is the most common type of cartilage. It covers the articular bone surfaces (where one or more bones meet at a joint). It also connects the ribs to the sternum and appears in the trachea, bronchi, and nasal septum.

Strong and rigid

Fibrous cartilage forms the symphysis pubis and the intervertebral disks. This type of cartilage is composed of small quantities of matrix and abundant fibrous elements. It's strong and rigid.

Elastic and resilient

Elastic cartilage is located in the auditory canal, external ear, and epiglottis. Large numbers of elastic fibers give this type of cartilage elasticity and resiliency.

Joints

Joints (articulations) are points of contact between two bones that hold the bones together. Many joints also allow flexibility and movement.

It's the joint

Joints can be classified by function (extent of movement) or by structure (what they're made of). The body has three major types of joints classified by function and three major types classified by structure.

Functional classification

By function, a joint may be classified as:

 synarthrosis (immovable)

Hey, I know a joint where we can get together.

amphiarthrosis (slightly movable)

 diarthrosis (freely movable).

Structural classification

By structure, a joint may be classified as:

 fibrous

 cartilaginous

 synovial.

Fibrous joints

In *fibrous joints*, the articular surfaces of the two bones are bound closely by fibrous connective tissue, and little movement is possible. Fibrous joints include sutures, syndesmoses (such as the radioulnar joints), and gomphoses (such as the dental alveolar joint).

Cartilaginous joints

In *cartilaginous joints* (also called amphiarthroses), cartilage connects one bone to another. Cartilaginous joints allow slight movement. They occur as:
• *synchondroses*, which are typically temporary joints in which the intervening hyaline cartilage converts to bone by adulthood — for example, the epiphyseal plates of long bones
• *symphyses*, which are joints with an intervening pad of fibrocartilage — for example, the symphysis pubis.

Synovial joints

The contiguous bony surfaces in the *synovial joints* are separated by a viscous, lubricating fluid—the *synovia* — and by cartilage. They're joined by ligaments lined with a synovia-producing membrane. Freely movable, synovial joints include most joints of the arms and legs.

Synovial joints are freely movable and include most joints of the arms and legs.

Other features of synovial joints include:
• a *joint cavity* — a potential space that separates the articulating surfaces of the two bones
• an *articular capsule* — a saclike envelope with outer layer that is lined with a vascular synovial membrane
• *reinforcing ligaments* — fibrous tissue that connects bones within the joint and reinforces the joint capsule.

Based on their structure and the type of movement they allow, synovial joints fall into various subdivisions — gliding, hinge, pivot, condylar, saddle, and ball-and-socket.

Let it glide

Gliding joints have flat or slightly curved articular surfaces and allow gliding movements. However, because they're bound by ligaments, they may not allow movement in all directions. Examples of gliding joints are the intertarsal and intercarpal joints of the hands and feet.

Here's a hinge

In *hinge joints*, a convex portion of one bone fits into a concave portion of another. The movement of a hinge joint resembles that of a metal hinge and is limited to flexion and extension. Hinge joints include the elbow and knee.

Pivot joints

A rounded portion of one bone in a *pivot joint* fits into a groove in another bone. Pivot joints allow only uniaxial rotation of the first bone around the second. An example of a pivot joint is the head of the radius, which rotates within a groove of the ulna.

Give condylar joints a hand

In *condylar joints*, an oval surface of one bone fits into a concavity in another bone. Condylar joints allow flexion, extension, abduction, adduction, and circumduction. Examples include the radiocarpal and metacarpophalangeal joints of the hand.

Saddle up

Saddle joints resemble condylar joints but allow greater freedom of movement. The only saddle joints in the body are the carpometacarpal joints of the thumb.

Ball-and-socket: These joints are hip

The *ball-and-socket joint* gets its name from the way its bones connect: The spherical head of one bone fits into a concave "socket" of another bone. The body's only ball-and-socket joints are the shoulder and hip joints.

> The ball-and-socket is the hippest joint around. It allows more freedom of movement than any other synovial joint.

Bursae

Bursae are small synovial fluid sacs that are located at friction points around joints between tendons, ligaments, and bones.

Stress reducers

Bursae act as cushions to decrease stress on adjacent structures. Examples of bursae include the subacromial bursa (located in the shoulder) and the prepatellar bursa (located in the knee).

Quick quiz

1. Which muscle type is considered voluntary (contraction is controlled at will)?
   A. Cardiac
   B. Visceral
   C. Skeletal

Answer: C. Skeletal muscle is voluntary, meaning it can be moved at will. The musculoskeletal system consists mostly of skeletal muscle.

2. Which of the following is *not* a function of cartilage?
   A. Protect body structures
   B. Support and shape body structures
   C. Cushion body structures

Answer: A. Cartilage is responsible for supporting, cushioning, and shaping body structures. Types of cartilage include fibrous, hyaline, and elastic.

3. The type of joint that permits free movement is classified as:
   A. synarthrosis.
   B. amphiarthrosis.
   C. diarthrosis.

Answer: C. Diarthroses include the ankles, wrists, knees, hips, and shoulders. These joints permit free movement.

4. The carpometacarpal joints of the thumb are classified as:
   A. pivot joints.
   B. saddle joints.
   C. hinge joints.

Answer: B. Saddle joints are similar to condylar joints. The only saddle joints in the body are the carpometacarpal joints of the thumbs.

Scoring

☆☆☆ If you answered all four questions correctly, outstanding. You've flexed your mental muscles and are ready to tackle the next system.

☆☆ If you answered two or three questions correctly, splendid. You've got the bare bones of this system down pat.

☆ If you answered fewer than two questions correctly, that's okay. Let's take our osteocytes and blast into the next chapter.

Nervous system

Just the facts

In this chapter, you'll learn:

♦ structure of the nervous system

♦ functions of the nervous system

♦ special sense organs and their function.

Introducing the nervous system

The nervous system coordinates all body functions, enabling a person to adapt to changes in internal and external environments. It has two main types of cells:
• neurons, the conducting cells
• neuroglia, the supportive cells.

Neuron: The basic unit

The *neuron* is the basic unit of the nervous system. This highly specialized conductor cell receives and transmits electrochemical nerve impulses. Delicate, threadlike nerve fibers called *axons* and *dendrites* extend from the central cell body and transmit signals. In a typical neuron, one axon and many dendrites extend from the cell body. (See *Parts of a neuron*, page 70.)

Axons

The *axon* conducts nerve impulses away from the cell body. A typical axon has terminal branches and is wrapped in a white, fatty, segmented covering called a *myelin sheath*. The myelin sheath is produced by *Schwann cells* — phagocytic cells separated by gaps called *nodes of Ranvier*.

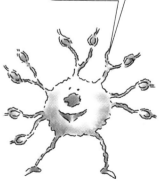

I'm a neuron, the fundamental unit of the nervous system.

Zoom in

Parts of a neuron

A typical neuron, like the one shown here, has one axon and many dendrites. A myelin sheath encloses the axon.

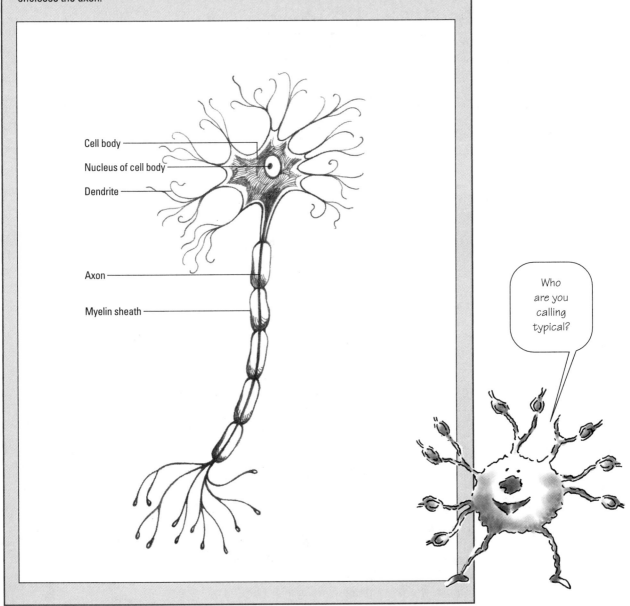

Cell body

Nucleus of cell body

Dendrite

Axon

Myelin sheath

Who are you calling typical?

Dendrites

Dendrites are short, thick, diffusely branched extensions of the cell body that receive impulses from other cells. Dendrites conduct impulses toward the cell body.

Sending the message

Neurons are responsible for *neurotransmission* — conduction of electrochemical impulses throughout the nervous system. Neuron activity may be provoked by:

mechanical stimuli, such as touch and pressure

thermal stimuli, such as heat and cold

 chemical stimuli, such as external chemicals or a chemical (such as histamine) released by the body. (See *How neurotransmission occurs*, page 72.)

> Stimuli such as heat produce neuron activity and start conduction of electrochemical impulses through the nervous system.

Neuroglia

Neuroglia (also called glial cells) are the supportive cells of the nervous system. They form roughly 40% of the brain's bulk. Four types of neuroglia exist:

• *Astroglia*, or *astrocytes*, exist throughout the nervous system. They supply nutrients to neurons and help them maintain their electrical potential as well as form part of the blood-brain barrier.

• *Ependymal cells* line the four small cavities in the brain, called ventricles, as well as the choroid plexuses. These cells help produce cerebrospinal fluid (CSF).

• *Microglia* are phagocytic cells that ingest and digest microorganisms and waste products from injured neurons.

• *Oligodendroglia* support and electrically insulate central nervous system (CNS) axons by forming protective myelin sheaths.

> Glial is derived from the Greek word for glue. Glial cells "glue" the neurons together.

Now I get it!

How neurotransmission occurs

Neurons receive and transmit stimuli by electrochemical messages. Dendrites on the neuron receive an impulse sent by other cells and conduct it toward the cell body. The axon then conducts the impulse away from the cell.

To stimulate or inhibit

When the impulse reaches the end of the axon, it stimulates synaptic vesicles in the presynap-tic axon terminal. A neurotransmitter substance is then released into the synaptic cleft between neurons. This substance diffuses across the synaptic cleft and binds to special receptors on the postsynaptic membrane. This stimulates or inhibits stimulation of the postsynaptic neuron.

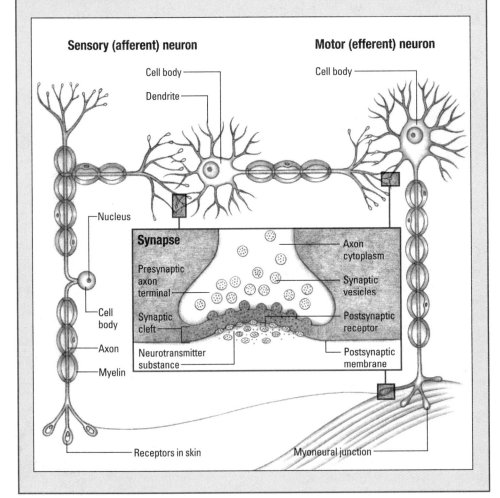

Central nervous system

The CNS includes the brain and the spinal cord. The CNS is encased by the bones of the skull and vertebral column and is protected by the cerebral spinal fluid and the meninges (dura mater, arachnoid, and pia mater).

Brain

The brain consists of the cerebrum, cerebellum, brain stem, diencephalon (thalamus and hypothalamus), limbic system, and reticular activating system.

Cerebrum

The *cerebrum* is the largest part of the brain. It houses the nerve center that controls sensory and motor activities and intelligence.

Touch of gray

The outer layer of the cerebrum, the *cerebral cortex*, consists of unmyelinated nerve fibers *(gray matter)*. The inner layer of the cerebrum consists of myelinated nerve fibers.

Steady as she goes

Within the cerebrum lies *white matter* (myelinated nerve fibers). *Basal ganglia*, which control motor coordination and steadiness, are found in white matter.

Bridging the hemispheres

The cerebrum has right and left hemispheres. A mass of nerve fibers known as the *corpus callosum* bridges the hemispheres, allowing communication between corresponding centers in each hemisphere. The rolling surface of the cerebrum is made up of *gyri* (convolutions) and *sulci* (creases or fissures).

> The corpus callosum allows communication between the two hemispheres.

The four lobes

Each cerebral hemisphere is divided into four lobes, based on anatomic landmarks and functional differences. The lobes are named for the cranial bones that lie over them — frontal, temporal, parietal, and occipital. (See *A close look at major brain structures.*)

Cerebellum

The *cerebellum* is the brain's second largest region. It lies behind and below the cerebrum. Like the cerebrum, it has two hemispheres. It also has an outer cortex of gray matter and an inner core of white matter. The cerebellum functions to maintain muscle tone, coordinate muscle movement, and control balance.

Brain stem

The *brain stem* lies immediately below the cerebrum, just in front of the cerebellum. It's continuous with the cerebrum above and with the spinal cord below.

It all stems from the brain stem

The brain stem consists of the *midbrain, pons,* and *medulla oblongata.* The brain stem relays messages between the parts of the nervous system. It has three main functions:
• It produces the rigid autonomic behaviors necessary for survival, such as increasing the heart rate and stimulating the adrenal medulla to produce epinephrine.
• It provides pathways for nerve fibers between higher and lower neural centers.
• It serves as the origin for 10 of the 12 pairs of cranial nerves.

Midbrain, pons, and medulla oblongata

The three parts of the brain stem provide two-way conduction between the spinal cord and brain. In addition, they perform the following functions:
• The *midbrain* is the reflex center for cranial nerves III and IV and mediates pupillary reflexes and eye movements.
• The *pons* helps regulate respirations. It connects the cerebellum with the cerebrum and links the midbrain to the medulla oblongata. It's also the reflex center for cranial nerves V through VIII. The pons mediates chewing, taste, saliva secretion, hearing, and equilibrium.

> Thanks to the brain stem, I can communicate with the rest of the nervous system.

Zoom in

A close look at major brain structures

This illustration shows the two largest structures of the brain — the cerebrum and cerebellum. Several fissures divide the cerebrum into hemispheres and lobes:

• The *fissure of Sylvius,* or the lateral sulcus, separates the temporal lobe from the frontal and parietal lobes.

• The *fissure of Rolando,* or the central sulcus, separates the frontal lobes from the parietal lobe.

• The *parieto-occipital fissure* separates the occipital lobe from the two parietal lobes.

To each lobe, a function

Each lobe has a particular function:

• The *frontal lobe* influences personality, judgment, abstract reasoning, social behavior, language expression, and movement (in the motor portion).

• The *temporal lobe* controls hearing, language comprehension, and storage and recall of memories (although memories are stored throughout the entire brain).

• The *parietal lobe* interprets and integrates sensations, including pain, temperature, and touch. It also interprets size, shape, distance, and texture. The parietal lobe of the nondominant hemisphere is especially important for awareness of body shape.

• The *occipital lobe* functions mainly to interpret visual stimuli.

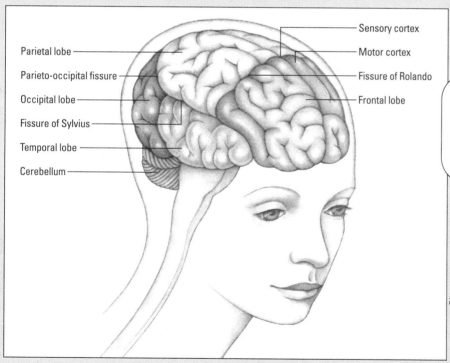

Parietal lobe

Parieto-occipital fissure

Occipital lobe

Fissure of Sylvius

Temporal lobe

Cerebellum

Sensory cortex

Motor cortex

Fissure of Rolando

Frontal lobe

I've got a charming personality, thanks to my frontal lobe.

• The *medulla oblongata* joins the spinal cord at the level of the *foramen magnum*, an opening in the occipital portion of the skull. It influences cardiac, respiratory, and vasomotor functions. It's the center for the vomiting, coughing, and hiccuping reflexes.

Diencephalon

The *diencephalon* is the part of the brain located between the cerebrum and the midbrain. It consists of the thalamus and hypothalamus, which lie beneath the surface of the cerebral hemispheres.

Thalamus at attention

The *thalamus* relays all sensory stimuli (except olfactory) as they ascend to the cerebral cortex. Its functions include primitive awareness of pain, screening of incoming stimuli, and focusing of attention.

Don't wake the hypothalamus

The *hypothalamus* controls or affects body temperature, appetite, water balance, pituitary secretions, emotions, and autonomic functions, including sleeping and waking cycles.

My thalamus acts as a relay station. Sensory impulses cross synapses in the thalamus on their way to the cerebral cortex.

Limbic system

The *limbic system* is a primitive brain area deep within the temporal lobe. In addition to initiating basic drives (hunger, aggression, and emotional and sexual arousal) the limbic system screens all sensory messages traveling to the cerebral cortex.

Reticular activating system

The *reticular activating system* (RAS) is a diffuse network of hyperexcitable neurons. It fans out from the brain stem through the cerebral cortex. After screening all incoming sensory information, the RAS channels it to appropriate areas of the brain for interpretation. It functions as the arousal, or alerting, system for the cerebral cortex, and its functioning is crucial for maintaining consciousness.

Memory jogger

To remember the function of the reticular activating system, focus on *activating*. When active, wakefulness is stimulated; when activity declines, a person falls asleep.

Oxygenating the brain

Four major arteries — two vertebral and two carotid — supply the brain with oxygenated blood.

Vertebral convergence

The two *vertebral arteries* (branches of the subclavians) converge to become the basilar artery. The *basilar artery* supplies oxygen to the posterior brain.

Two carotids diverged in the brain...

The common carotids branch into the two internal carotids, which divide further to supply oxygen to the anterior brain and the middle brain. These arteries interconnect through the circle of Willis, an anastomosis at the base of the brain. The circle of Willis ensures that oxygen is continually circulated to the brain despite interruption of any of the brain's major vessels. (See *Arteries of the brain*, page 78.)

> Because of the circle of Willis, blood has two paths to the brain, ensuring a continuous oxygen supply.

Spinal cord

The *spinal cord* is a cylindrical structure in the vertebral canal that extends from the foramen magnum at the base of the skull to the upper lumbar region of the vertebral column.

Getting on my spinal nerves

The *spinal nerves* arise from the cord. At the cord's inferior end, nerve roots cluster in the *cauda equina*.

What's the matter in the spinal cord?

Within the spinal cord, the H-shaped mass of gray matter is divided into *horns*. Horns consist mainly of neuron cell bodies. Cell bodies in the two dorsal (posterior) horns primarily relay sensations; those in the two ventral (anterior) horns play a part in voluntary and reflex motor activity. White matter surrounds the horns. This white matter consists of myelinated nerve fibers grouped in vertical columns, or *tracts*. In other words, all axons that compose one tract serve one general function, such as touch, movement, pain, and pressure. (See *A look inside the spinal cord*, page 79.)

Zoom in

Arteries of the brain

This illustration shows the inferior surface of the brain. The anterior and posterior arteries join with smaller arteries to form the circle of Willis.

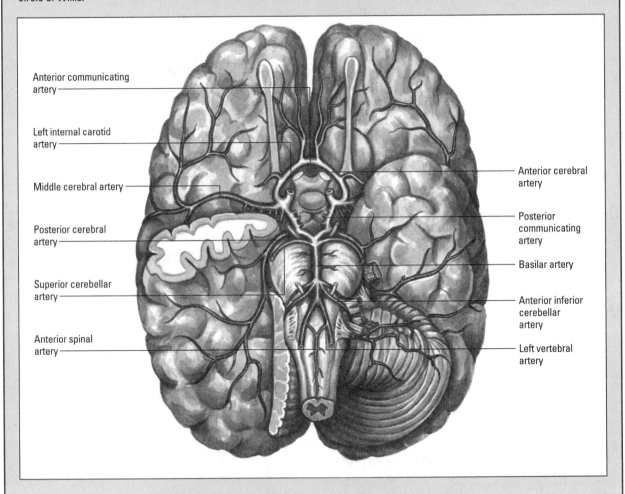

Anterior communicating artery

Left internal carotid artery

Middle cerebral artery

Posterior cerebral artery

Superior cerebellar artery

Anterior spinal artery

Anterior cerebral artery

Posterior communicating artery

Basilar artery

Anterior inferior cerebellar artery

Left vertebral artery

Sensory pathways

Sensory impulses travel via the *afferent* (sensory, or ascending) *neural pathways* to the *sensory cortex* in the parietal lobe of the brain. This is where the impulses are

Zoom in

A look inside the spinal cord

This cross section of the spinal cord shows an H-shaped mass of gray matter divided into horns, which consist primarily of neuron cell bodies. Cell bodies in the posterior, or dorsal, horn primarily relay information. Cell bodies in the anterior, or ventral, horn are needed for voluntary or reflex motor activity.

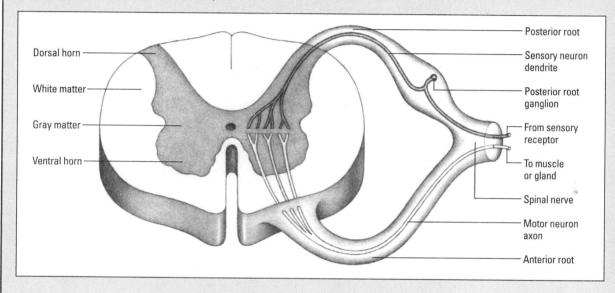

interpreted. These impulses use two major pathways: the dorsal horn and the ganglia.

Enter the dorsal horn

Pain and temperature sensations enter the spinal cord through the *dorsal horn*. After immediately crossing over to the opposite side of the cord, these impulses then travel to the thalamus via the spinothalamic tract.

Enter the ganglia

Touch, pressure, and vibration sensations enter the cord via relay stations called *ganglia*. Ganglia are knotlike masses of nerve cell bodies on the dorsal roots of spinal nerves. Impulses travel up the cord in the dorsal column to the medulla, where they cross to the opposite side and enter the thalamus. The thalamus relays all incoming sensory impulses (except olfactory impulses) to the sensory cortex for interpretation.

Motor pathways

Motor impulses travel from the brain to the muscles via the *efferent* (motor, or descending) *neural pathways*. Motor impulses originate in the *motor cortex* of the frontal lobe and reach the lower motor neurons of the peripheral nervous system via upper motor neurons.

Upper motor neurons originate in the brain and form two major systems:

 the pyramidal system

 the extrapyramidal system.

Skilled movements of skeletal muscle

The *pyramidal system* (corticospinal tract) is responsible for fine, skilled movements of skeletal muscle. Impulses in this system travel from the motor cortex through the internal capsule to the medulla. At the medulla they cross to the opposite side and continue down the spinal cord.

Get your gross motor running

The *extrapyramidal system* (extracorticospinal tract) controls gross motor movements. Impulses originate in the premotor area of the frontal lobes and travel to the pons. At the pons the impulses cross to the opposite side. Then the impulses travel down the spinal cord to the anterior horns, where they're relayed to the lower motor neurons. These neurons, in turn, carry the impulses to the muscles. (See *Major neural pathways*.)

Reflex responses

Reflex responses occur automatically, without any brain involvement, to protect the body. Spinal nerves, which have both sensory and motor portions, mediate *deep tendon reflexes* (involuntary contractions of a muscle after brief stretching caused by tendon percussion), *superficial reflexes* (withdrawal reflexes elicited by noxious or tactile stimulation of the skin, cornea, or mucous membranes) and, in infants, *primitive reflexes*.

The pyramidal system is responsible for fine, skilled movements.

Now I get it!

Major neural pathways

Sensory and motor impulses travel through different pathways to the brain for interpretation.

Sensory pathways
Sensory impulses travel through two major sensory (afferent, or ascending) pathways to the sensory cortex in the cerebrum.

Motor pathways
Motor impulses travel from the motor cortex in the cerebrum to the muscles via motor (efferent, or descending) pathways.

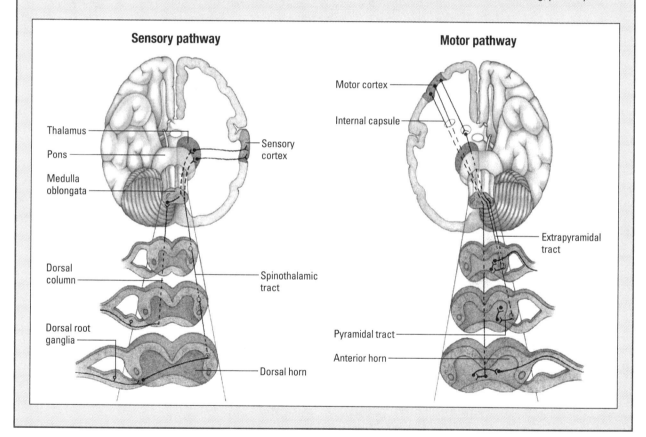

Deep we go

Deep tendon reflexes include reflex responses of the biceps, triceps, brachioradialis, patellar, and Achilles tendons. (See *Eliciting deep tendon reflexes*, page 82.)
• *Biceps reflex* contracts the biceps muscle and forces flexion of the forearm.

Peak technique

Eliciting deep tendon reflexes

There are five deep tendon reflexes. The methods of eliciting these reflexes are described below.

Biceps reflex

Placing the thumb or index finger over the biceps tendon and the remaining fingers loosely over the triceps muscle, strike the thumb or index finger over the biceps tendon with the pointed end of the reflex hammer. Watch and feel for the contraction of the biceps muscle and flexion of the forearm.

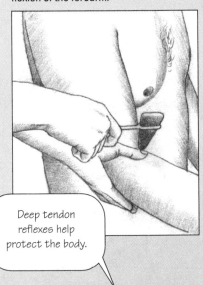

Triceps reflex

Strike the triceps tendon about 2″ (5 cm) above the olecranon process on the extensor surface of the upper arm. Watch for contraction of the triceps muscle and extension of the forearm.

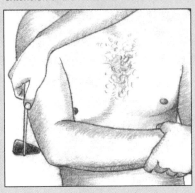

Brachioradialis reflex

Strike the radius about 1″ to 2″ (2.5 to 5 cm) above the wrist and watch for supination of the hand and flexion of the forearm at the elbow.

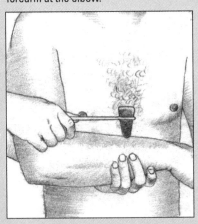

Patellar reflex

Strike the patellar tendon just below the patella and look for contraction of the quadriceps muscle in the thigh with extension of the leg.

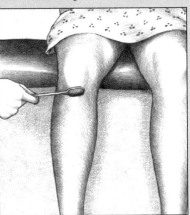

Achilles reflex

With the foot flexed and supporting the plantar surface, strike the Achilles tendon. Watch for plantar flexion of the foot at the ankle.

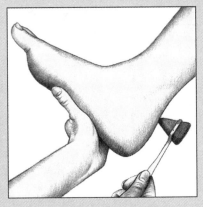

> Deep tendon reflexes help protect the body.

• *Triceps reflex* contracts the triceps muscle and forces extension of the forearm.
• *Brachioradialis reflex* causes supination of the hand and flexion of the forearm at the elbow.
• *Patellar reflex* forces contraction of the quadriceps muscle in the thigh with extension of the leg.
• *Achilles reflex* forces plantar flexion of the foot at the ankle.

Rising to the superficial

Superficial reflexes are reflexes of the skin and mucous membranes. Successive attempts to stimulate these reflexes provoke increasingly limited responses. Superficial reflexes include the following:
• *Plantar flexion* of the toes occurs when the lateral sole of an adult's foot is stroked from heel to great toe with a tongue blade.
• *Babinski's response* causes upward movement of the great toe and fanning of the little toes in children under age 2 in response to stimulation of the outer margin of the sole of the foot.
• In men, the *cremasteric reflex* is stimulated by stroking the inner thigh. This forces the contraction of the cremaster muscle and elevation of the testicle on the side of the stimulus.
• *Abdominal reflex* is induced by stroking the sides of the abdomen above and below the umbilicus, moving from the periphery toward the midline. Movement of the umbilicus toward the stimulus is normal.

Let's get primitive

Primitive reflexes are abnormal in adults but normal in infants, whose central nervous systems are immature. As the neurologic system matures, these reflexes disappear. The primitive reflexes are *grasping*, *sucking*, and *glabella:*
• The application of gentle pressure to an infant's palm results in grasping.
• An infantile sucking reflex to ingest milk is a primitive response to oral stimuli.
• The glabella reflex is elicited by repeatedly tapping the bridge of the infant's nose. The normal response is persistent blinking.

Achilles was a warrior in Homer's epic, the *Iliad.* He could only be killed by an arrow piercing the back of his heel. Therefore, the tendon located in this area is called the Achilles tendon.

Protective structures

The brain and spinal cord are protected from shock and infection by the bony skull and vertebrae, CSF, and three membranes: the dura mater, arachnoid membrane, and pia mater.

Dura mater

The *dura mater* is tough, fibrous, leatherlike tissue composed of two layers — the endosteal dura and meningeal dura.

The unending endosteal dura

The *endosteal dura* forms the periosteum of the skull and is continuous with the lining of the vertebral canal.

The durable meningeal dura

The *meningeal dura* is a thick membrane that covers the brain, dipping between the brain tissue and providing support and protection.

Arachnoid membrane

The *arachnoid membrane* is a thin, fibrous membrane that hugs the brain and spinal cord, though not as precisely as the pia mater.

Pia mater

The *pia mater* is a continuous, delicate layer of connective tissue that covers and contours the spinal tissue and brain.

The spaces in between

The *subdural space* lies between the dura mater and the arachnoid membrane. The *subarachnoid space* lies between the pia mater and the arachnoid membrane. Within the subarachnoid space and the brain's four ventricles is CSF, a fluid composed of water and traces of organic materials (especially protein), glucose, and minerals. This fluid protects the brain and spinal tissue from jolts and blows.

If the brain can't send a message to a patient's leg after a severe spinal cord injury, a stimulus can still cause the knee-jerk reflex...

...as long as the spinal cord remains intact at the level of this reflex.

Peripheral nervous system

The peripheral nervous system consists of the cranial nerves, spinal nerves, and autonomic nervous system (ANS).

Cranial nerves

The 12 pairs of cranial nerves transmit motor or sensory messages (or both) primarily between the brain or brain stem and the head and neck. All cranial nerves except the olfactory and optic nerves exit from the midbrain, pons, or medulla oblongata of the brain stem. (See *Exit points for the cranial nerves*, page 86.)

Spinal nerves

Each of the 31 pairs of spinal nerves is named for the vertebra immediately below the nerve's exit point from the spinal cord. From top to bottom they're designated as C1 through S5 and the coccygeal nerve. Each spinal nerve consists of *afferent* (sensory) and *efferent* (motor) neurons, which carry messages to and from particular body regions, called *dermatomes.*

Autonomic nervous system

The vast ANS *innervates* (supplies nerves to) all internal organs. Sometimes known as *visceral efferent nerves,* the nerves of the ANS carry messages to the viscera from the brain stem and neuroendocrine regulatory centers. The ANS has two major subdivisions: the *sympathetic* (thoracolumbar) nervous system and *parasympathetic* (craniosacral) nervous system.

When one system stimulates certain smooth muscles to contract or a gland to secrete, the other system inhibits that action. Through this dual innervation, the two divisions counterbalance each other's activities to keep body systems running smoothly.

Exit points for the cranial nerves

As this illustration reveals, 10 of the 12 pairs of cranial nerves (CN) exit from the brain stem. The remaining two pairs — the olfactory and optic nerves — exit from the forebrain.

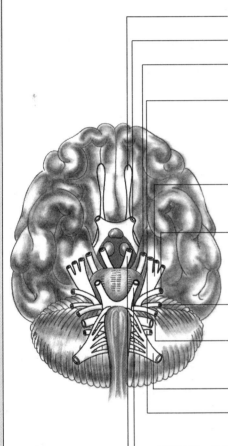

Olfactory (CN I). *Sensory:* smell

Optic (CN II). *Sensory:* vision

Trochlear (CN IV). *Motor:* extraocular eye movement (inferior medial)

Vagus (CN X). *Motor:* movement of palate, swallowing, gag reflex, activity of the thoracic and abdominal viscera, such as heart rate and peristalsis. *Sensory:* sensations of throat, larynx, and thoracic and abdominal viscera (heart, lungs, bronchi, and GI tract)

Trigeminal (CN V). *Sensory:* transmitting stimuli from face and head, corneal reflex. *Motor:* chewing, biting, and lateral jaw movements

Facial (CN VII). *Sensory:* taste receptors (anterior two-thirds of tongue). *Motor:* facial muscle movement, including muscles of expression (those in the forehead and around the eyes and mouth)

Acoustic (CN VIII). *Sensory:* hearing, sense of balance

Glossopharyngeal (CN IX). *Motor:* swallowing movements. *Sensory:* sensations of throat, taste receptors (posterior one-third of tongue)

Hypoglossal (CN XII). *Motor:* tongue movement

Spinal accessory (CN XI). *Motor:* shoulder movement, head rotation

Abducens (CN VI). *Motor:* extraocular eye movement (lateral)

Oculomotor (CN III). *Motor:* extraocular eye movement (superior, medial, and inferior lateral), pupillary constriction, upper eyelid elevation

Sympathetic nervous system

Sympathetic nerves called *preganglionic neurons* exit the spinal cord between the levels of the first thoracic and second lumbar vertebrae.

Ganglia branch out

When they leave the spinal cord, these nerves enter small ganglia near the cord. The ganglia form a chain that spreads the impulse to *postganglionic neurons*. Postganglionic neurons reach many organs and glands and can produce widespread, generalized physiologic responses. These responses include:
- vasoconstriction
- elevated blood pressure
- enhanced blood flow to skeletal muscles
- increased heart rate and contractility
- increased respiratory rate
- smooth-muscle relaxation of the bronchioles, GI tract, and urinary tract
- sphincter contraction
- pupillary dilation and ciliary muscle relaxation
- increased sweat gland secretion
- reduced pancreatic secretion.

Parasympathetic nervous system

Fibers of the parasympathetic nervous system leave the CNS by way of the cranial nerves from the midbrain and medulla and the spinal nerves between the second and fourth sacral vertebrae (S2 to S4).

Leaving the CNS

After leaving the CNS, the long preganglionic fiber of each parasympathetic nerve travels to a ganglion near a particular organ or gland. The short postganglionic fiber enters the organ or gland. This creates a more specific response involving only one organ or gland.

Such a response might be:
- reductions in heart rate, contractility, and conduction velocity
- bronchial smooth-muscle constriction
- increased GI tract tone and peristalsis, with sphincter relaxation
- increased bladder tone and urinary system sphincter relaxation

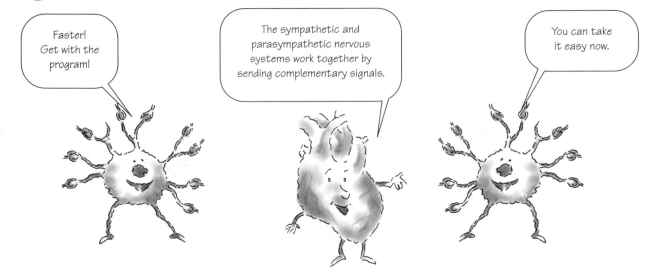

- vasodilation of external genitalia, causing erection
- pupil constriction
- increased pancreatic, salivary, and lacrimal secretions.

Special sense organs

Sensory stimulation allows the body to interact with the environment. The distal ends of the dendrites of sensory neurons serve as sensory receptors, sending messages to the brain. The brain also receives stimulation from the special sense organs — the eyes, the ears, and the gustatory and olfactory organs.

Extraocular eye structure

The *eye* is the organ of vision and contains about 70% of the body's sensory receptors. Although the eye measures about 1″ (2.5 cm) in diameter, only its anterior surface is visible.

Extraocular muscles hold the eyes in place and control their movement. Their coordinated action keeps both eyes parallel and creates binocular vision. These muscles have mutually antagonistic actions: As one muscle contracts, its opposing muscle relaxes.

Extraocular structures include the eyelids, conjunctivae, and lacrimal apparatus. Together with the extraocular muscles, these structures support and protect the eyeball.

Eyelids

The *eyelids* (also called the palpebrae) are loose folds of skin that cover the anterior portion of the eye. The lid margins contain hair follicles, which contain eyelashes and sebaceous glands.

In the blink of an eye

When open, the upper eyelid extends beyond the *limbus* (the junction of the cornea and the sclera) and covers a small portion of the iris.

The eyelids contain three types of glands:
- *meibomian glands* — sebaceous glands that secrete sebum, an oily substance that keeps the eye lubricated
- *glands of Zeis* — modified sebaceous glands connected to the follicles of the eyelashes
- *Moll's glands* — ordinary sweat glands.

When closed, the upper and lower eyelids cover the eye completely.

Conjunctivae

Conjunctivae are thin mucous membranes that line the inner surface of each eyelid and the anterior portion of the sclera. Conjunctivae guard the eye from invasion by foreign matter. The *palpebral conjunctiva* — the portion that lines the inner surface of the eyelids — appears shiny pink or red. The *bulbar* (or ocular) *conjunctiva*, which joins the palpebral portion and covers the exposed part of the sclera, contains many small, normally visible blood vessels.

Lacrimal apparatus

The structures of the *lacrimal apparatus* (lacrimal glands, punctum, lacrimal sac, and nasolacrimal duct) lubricate and protect the cornea and conjunctivae by producing and absorbing tears. Tears keep the cornea and conjunctivae moist. Tears also contain *lysozyme*, an enzyme that protects against bacterial invasion.

Remember that with extraocular muscles, as one muscle contracts, its opposing muscle relaxes.

The eyelid protects the eye, regulates the entrance of light, and distributes tears over the eye.

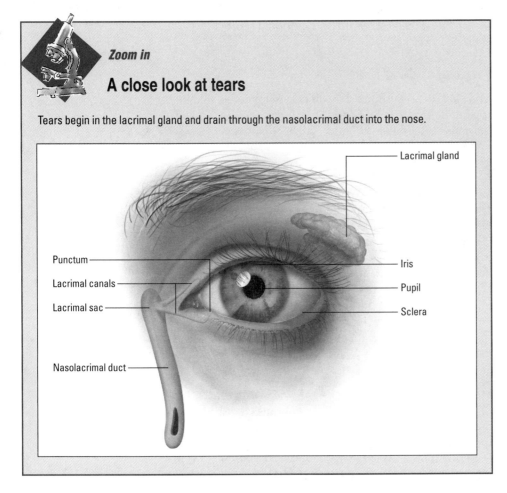

Zoom in

A close look at tears

Tears begin in the lacrimal gland and drain through the nasolacrimal duct into the nose.

Lacrimal gland

Punctum

Lacrimal canals

Lacrimal sac

Nasolacrimal duct

Iris

Pupil

Sclera

These darn tears will be the death of me!

Cry me a river

As the eyelids blink, they direct the flow of tears from the lacrimal ducts to the *inner canthus*, the medial angle between the eyelids. Tears pool at the inner canthus and drain through the *punctum*, a tiny opening. From there they flow through the *lacrimal canals* into the lacrimal sac. Lastly, they drain through the nasolacrimal duct and into the nose. (See *A close look at tears.*)

Intraocular eye structures

Intraocular structures within the eyeball are directly involved with vision. (See *Looking at intraocular structures.*)

Zoom in

Looking at intraocular structures

Some intraocular structures, such as the sclera, cornea, iris, pupil, and anterior chamber, are visible to the naked eye. Others, such as the retina, are visible only with an ophthalmoscope. These illustrations show the major structures within the eye.

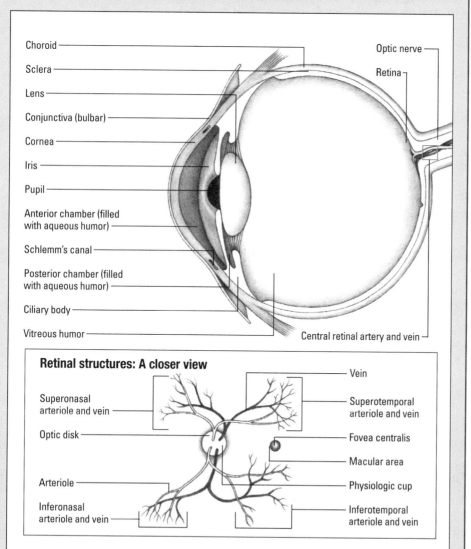

Choroid

Sclera

Lens

Conjunctiva (bulbar)

Cornea

Iris

Pupil

Anterior chamber (filled with aqueous humor)

Schlemm's canal

Posterior chamber (filled with aqueous humor)

Ciliary body

Vitreous humor

Optic nerve

Retina

Central retinal artery and vein

Retinal structures: A closer view

Superonasal arteriole and vein

Optic disk

Arteriole

Inferonasal arteriole and vein

Vein

Superotemporal arteriole and vein

Fovea centralis

Macular area

Physiologic cup

Inferotemporal arteriole and vein

Want to see the sclera, cornea, iris, or pupil? Just look in the mirror.

The sclera, cornea, iris, pupil, anterior chamber, aqueous humor, lens, ciliary body, and posterior chamber are found in the anterior segment. The vitreous humor, posterior sclera, choroid, and retina are found in the posterior segment.

Sclera and cornea

The white *sclera* coats four-fifths of the outside of the eyeball, maintaining its size and form. The *cornea* is continuous with the sclera at the limbus, revealing the pupil and iris. A smooth, transparent tissue, the cornea has no blood supply. The corneal epithelium merges with the bulbar conjunctiva at the limbus. The cornea is highly sensitive to touch and is kept moist by tears.

Iris and pupil

The *iris* is a circular contractile disk that contains smooth and radial muscles. It has an opening in the center for the *pupil*. Eye color depends on the amount of pigment in the endothelial layers of the iris. Pupil size is controlled by involuntary dilatory and sphincter muscles in the posterior region of the iris that regulate light entry.

Anterior chamber and aqueous humor

The *anterior chamber* is a cavity bounded in front by the cornea and behind by the lens and iris. It's filled with a clear, watery fluid called *aqueous humor*.

Lens

The *lens* is situated directly behind the iris at the pupillary opening. The lens acts like a camera lens, refracting and focusing light onto the retina. It's composed of transparent fibers in an elastic membrane called the *lens capsule*.

Ciliary body

The *ciliary body* (three muscles along with the iris that make up the anterior part of the vascular uveal tract) controls the lens thickness. Together with the coordinated action of muscles

> Moisture plays an important role in the eye. For example, tears protect the cornea by keeping it moist.

in the iris, the ciliary body regulates the light focused through the lens onto the retina.

Posterior chamber

The *posterior chamber* is a small space directly posterior to the iris but anterior to the lens. It's filled with aqueous humor.

Vitreous humor

The *vitreous humor* consists of a thick, gelatinous material that fills the space behind the lens. There, it maintains placement of the retina and the spherical shape of the eyeball.

Posterior sclera and choroid

The *posterior sclera* is a white, opaque, fibrous layer that covers the posterior segment of the eyeball. It continues back to the dural sheath, covering the optic nerve. The *choroid* lies beneath the posterior sclera. It contains many small arteries and veins.

Retina

The *retina* is the innermost coat of the eyeball. It receives visual stimuli and sends them to the brain. Each of the four sets of retinal vessels contains a transparent arteriole and vein as well as the optic disk, the physiologic cup, rods and cones, and the macula.

The optimal optic disk

Arterioles and veins become progressively thinner as they leave the optic disk. The *optic disk* is a well-defined, 1.5-mm round or oval area on the retina. Creamy yellow to pink in color, the optic disk allows the optic nerve to enter the retina at a point called the *nerve head*. A whitish to grayish crescent of scleral tissue may be present on the lateral side of the disk.

The cup within the disk

The *physiologic cup* is a light-colored depression within the optic disk on the temporal side. It covers one-third of the center of the disk.

> The posterior segment of the eye has three layers: fibrous, vascular, and sensory.

Rods and cones

Photoreceptor neurons called *rods* and *cones* compose the visual receptors of the retina. These receptors are responsible for vision.

Dark macula

The *macula* is lateral to the optic disk. It's slightly darker than the rest of the retina and without visible retinal vessels. A slight depression in the center of the macula, known as the *fovea centralis*, contains the heaviest concentration of cones and is a main receptor for vision and color.

Vision pathway

Intraocular structures perceive and form images and then send them to the brain for interpretation. To interpret these images properly, the brain relies on structures along the vision pathway. The *vision pathway* uses the optic nerve, optic chiasm, and retina to create the proper visual fields.

Optic chiasm and optic tracts

In the *optic chiasm*, fibers from the nasal aspects of both retinas cross to the opposite sides, and fibers from the temporal portions remain uncrossed. These crossed and uncrossed fibers form the *optic tracts*. Injury to one of the optic nerves can cause blindness in the corresponding eye. An injury or lesion in the optic chiasm can cause partial vision loss (for example, loss of the two temporal visual fields).

Focusing on the fovea centralis

Image formation begins when eye structures refract light rays from an object. Normally, the cornea, aqueous humor, lens, and vitreous humor refract light rays from an object, focusing them on the fovea centralis, where an inverted and reversed image clearly forms. Within the retina, rods and cones turn the projected image into an impulse and transmit it to the optic nerve.

Follow the tracts to the cerebral cortex

The impulse travels to the optic chiasm, where the two optic nerves unite, split again into two optic tracts, and

> The optic disk has no light receptors and is therefore a "blind spot." But I compensate so that there's usually no apparent gap in what is seen.

then continue into the optic section of the cerebral cortex. There, the inverted and reversed image on the retina is processed by the brain to create an image as it truly appears in the field of vision.

External ear structures

The *ears* are the organs of hearing. They also maintain the body's equilibrium. The ear is divided into three main parts: external, middle, and inner. (See *Ear structures*, page 96.)

The *external ear* consists of the *auricle* (pinna) and the *external auditory canal*. The *mastoid process* isn't part of the external ear but is an important bony landmark behind the lower part of the auricle.

Middle ear structures

The *middle ear* is also called the *tympanic cavity*. It's an air-filled cavity within the hard portion of the temporal bone. The tympanic cavity is lined with mucosa. It's bounded distally by the tympanic membrane and medially by the oval and round windows.

Tympanic membrane

The *tympanic membrane* consists of layers of skin, fibrous tissue, and a mucous membrane. It transmits sound vibrations to the internal ear.

Eustachian tube

The *eustachian*, or auditory, *tube* extends downward, forward, and inward from the middle ear cavity to the nasopharynx. It has a useful function: It makes possible the equalization of pressure against inner and outer surfaces of the tympanic membrane and prevents rupture.

Oval window

The *oval window* (fenestra ovalis) is an opening in the wall between the middle and inner ears into which part of the *stapes* (a tiny bone of the middle ear) fits. It transmits vibrations to the inner ear.

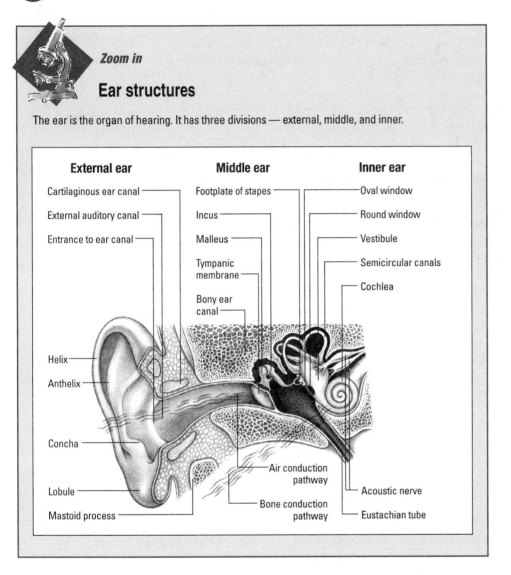

Zoom in

Ear structures

The ear is the organ of hearing. It has three divisions — external, middle, and inner.

External ear

Cartilaginous ear canal

External auditory canal

Entrance to ear canal

Helix

Anthelix

Concha

Lobule

Mastoid process

Middle ear

Footplate of stapes

Incus

Malleus

Tympanic membrane

Bony ear canal

Air conduction pathway

Bone conduction pathway

Inner ear

Oval window

Round window

Vestibule

Semicircular canals

Cochlea

Acoustic nerve

Eustachian tube

Round window

The *round window* (fenestra cochleae) is another opening in the same wall. It's enclosed by the secondary tympanic membrane. Like the oval window, the round window transmits vibrations to the inner ear.

Small bones

The middle ear contains three small bones, called *ossicles*, that conduct vibratory motion of the tympanum to the oval window. The ossicles are:

- the *malleus* (hammer), which attaches to the tympanic membrane and transfers sound to the incus
- the *incus* (anvil), which articulates the malleus and the stapes and carries vibration to the stapes
- the *stapes* (stirrup), which connects vibratory motion from the incus to the oval window.

Inner ear structures

In the inner ear, vibration excites receptor nerve endings. A bony labyrinth and a membranous labyrinth combine to form the inner ear. The inner ear contains the vestibule, cochlea, and semicircular canals.

Vestibule

The *vestibule* is located posterior to the cochlea and anterior to the semicircular canals. It serves as the entrance to the inner ear. It houses two membranous sacs, the *saccule* and *utricle*. Suspended in a fluid called *perilymph*, the saccule and utricle sense gravity changes and linear and angular acceleration.

Cochlea

The *cochlea*, a bony, spiraling cone, extends from the anterior part of the vestibule. Within it lies the *cochlear duct*, a triangular, membranous structure that houses the *organ of Corti*. The receptor organ for hearing, the organ of Corti transmits sound to the cochlear branch of the acoustic nerve (cranial nerve VIII).

Semicircular canals

The three *semicircular canals* project from the posterior aspect of the vestibule. Each canal is oriented in one of three planes: superior, posterior, and lateral. The *semicircular duct* traverses the canals and connects with the utricle anteriorly. The *crista ampullaris* sits at the end of each canal and contains hair cells and support cells. It's stimulated by sudden movements or changes in the rate or direction of movement.

Hearing pathways

For hearing to occur, sound waves travel through the ear by two pathways — air conduction and bone conduction. *Air conduction* occurs when sound waves travel in the air through the external and middle ear to the inner ear. *Bone conduction* occurs when sound waves travel through bone to the inner ear.

Interpreting the vibrations, man

Vibrations transmitted through air and bone stimulate nerve impulses in the inner ear. The cochlear branch of the acoustic nerve transmits these vibrations to the auditory area of the cerebral cortex. The cerebral cortex then interprets the sound.

> For hearing to occur, sound waves travel through the ear by two pathways — air conduction and bone conduction.

Nose and mouth

The *nose* is the sense organ for smell. The mucosal epithelium that lines the uppermost portion of the nasal cavity houses receptors for fibers of the olfactory nerve (cranial nerve I).

Good old olfactory

These receptors, called *olfactory* (smell) *receptors*, consist of hair cells. These hair cells are highly sensitive but easily fatigued. They're stimulated by the slightest odors but stop sensing even the strongest smells after a short time.

Slip of the tongue

The tongue and the roof of the mouth contain most of the receptors for the taste nerve fibers (located in branches of cranial nerves VII and IX). Called *taste buds*, these receptors are stimulated by chemicals. They respond to four taste sensations: sweet, sour, bitter, and salty. All the other flavors a person senses result from a combination of olfactory-receptor and taste-bud stimulation.

Quick quiz

1. The components of the central nervous system include:

    A.   the spinal cord and cranial nerves.

    B.   the brain and spinal cord.

    C.   the sympathetic and parasympathetic nervous systems.

Answer: B. The two main divisions of the nervous system are the CNS, which includes the brain and spinal cord, and the peripheral nervous system, which consists of the cranial nerves, spinal nerves, and ANS.

2. The brain is protected from shock and infection by:

    A.   bones, the meninges, and CSF.

    B.   gray matter, bones, and the primitive structures.

    C.   the blood-brain barrier, CSF, and white matter.

Answer: A. Bones (the skull and vertebral column), the meninges, and CSF protect the brain from shock and infection.

3. The visual receptors of the retina are composed of:

    A.   rods and cones.

    B.   the optic disk and optic nerve.

    C.   vitreous humor and aqueous humor.

Answer: A. Photoreceptor neurons called rods and cones compose the visual receptors of the retina.

4. The external ear consists of:

    A.   the vestibule, cochlea, and semicircular ducts.

    B.   the tympanic membrane, oval window, and round window.

    C.   the auricle and external auditory canal.

Answer: C. The auricle and external auditory canal are part of the external ear.

5. The cranial nerves transmit motor and sensory messages between the:
   A.   spine and body dermatomes.
   B.   brain and the head and neck.
   C.   viscera and the brain.
Answer: B. The 12 pairs of cranial nerves transmit motor (efferent) and sensory (afferent) messages between the brain or brainstem and the head and neck.

Scoring

☆☆☆   If you answered all five questions correctly, stupendous! You're definitely brainy.

☆☆   If you answered three or four questions correctly, remarkable. Your gray matter is working at lightning speed.

☆   If you answered fewer than three questions correctly, don't get nervous. There are 11 more quick quizzes to go.

Endocrine system

Just the facts

In this chapter, you'll learn:

♦ how the glands of the endocrine system function

♦ how hormones are released and transported

♦ role receptors play in the influence of hormones on cells.

In the endocrine system

The three major components of the endocrine system are:
- *glands* — specialized cell clusters or organs
- *hormones* — chemical substances secreted by glands in response to stimulation
- *receptors* — protein molecules that trigger specific physiologic changes in a target cell in response to hormonal stimulation.

Glands

The major glands of the endocrine system are:
- pituitary gland
- thyroid gland
- parathyroid glands
- adrenal glands
- pancreas
- thymus
- pineal gland
- gonads (ovaries and testes). (See *Components of the endocrine system*, page 102.)

Along with the nervous system, the endocrine system regulates and integrates the body's metabolic activities.

Body shop

Components of the endocrine system

Endocrine glands secrete hormones directly into the bloodstream to regulate body function. This illustration shows the location of the major endocrine glands (except the gonads).

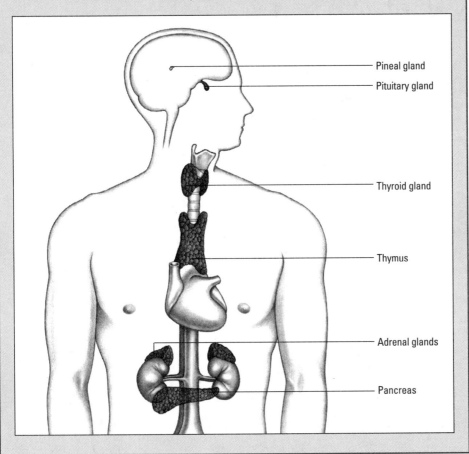

Pineal gland

Pituitary gland

Thyroid gland

Thymus

Adrenal glands

Pancreas

The pituitary gland: Small but mighty

The *pituitary gland* (also called the hypophysis or master gland) rests in the *sella turcica*, a depression in the sphenoid bone at the base of the brain.

This pea-sized gland connects with the hypothalamus via the infundibulum, from which it receives chemical

Now I get it!

How the hypothalamus affects endocrine activities

The *hypothalamus* is the integrative center for the endocrine and autonomic (involuntary) nervous systems. It helps control some endocrine glands by neural and hormonal stimulation.

Take the path
Neural pathways connect the hypothalamus to the posterior pituitary gland. These hypothalamic neurons stimulate the posterior pituitary gland to secrete two effector hormones — antidiuretic hormone (ADH) and oxytocin — which are stored in the posterior pituitary.

What happens after secretion
When ADH is secreted, the body retains water. Oxytocin stimulates uterine contractions during labor and milk secretion in lactating women.

The regulator
The hypothalamus also produces many other inhibiting and stimulating hormones and other factors, which it uses to regulate functions of the anterior pituitary.

> The hypothalamus is the integrative center for the endocrine and autonomic nervous systems.

and nervous stimulation. (See *How the hypothalamus affects endocrine activities.*)

Anterior pituitary

The pituitary has two main regions. The *anterior pituitary* (adenohypophysis) is the larger region. It produces at least six hormones:
- growth hormone (GH), or somatotropin
- thyroid-stimulating hormone (TSH), or thyrotropin
- corticotropin (ACTH)
- follicle-stimulating hormone (FSH)
- luteinizing hormone (LH)
- prolactin.

Posterior pituitary

The *posterior pituitary* makes up about 25% of the gland. It serves as a storage area for antidiuretic hormone (ADH), or vasopressin, and oxytocin, which are produced by the hypothalamus.

Thyroid gland

The *thyroid* lies directly below the larynx, partially in front of the trachea. Its two lateral lobes — one on either side of the trachea — join with a narrow tissue bridge, called the *isthmus*, to give the gland its butterfly shape.

Two lobes that function as one

The two lobes of the thyroid function as one unit to produce the hormones *triiodothyronine (T_3)*, *thyroxine (T_4)*, and *calcitonin*.

T_3 and T_4 equal thyroid hormone

T_3 and T_4 are collectively referred to as *thyroid hormone*. Thyroid hormone is the body's major metabolic hormone. It regulates metabolism by speeding cellular respiration.

Calcitonin is maintain'

Calcitonin maintains the blood calcium level. It does this by inhibiting the release of calcium from bone. Secretion of calcitonin is controlled by the calcium concentration of the fluid surrounding the thyroid cells.

Parathyroid glands

The *parathyroid glands* are the body's smallest known endocrine glands. These glands are embedded on the posterior surface of the thyroid, one in each corner.

Parathyroid hormone, a.k.a. PTH

Working together as a single gland, the parathyroid glands produce *parathyroid hormone* (PTH). The main function of PTH is to help regulate the blood's calcium

Whew! I couldn't do this without thyroid hormone.

Right, we'd be couch potatoes without T_3 and T_4.

balance. This hormone adjusts the rate at which calcium and magnesium ions are removed from urine. PTH also increases the movement of phosphate ions from the blood to urine for excretion.

Adrenal glands

The two *adrenal glands* each lie on top of a kidney. These almond-shaped glands contain two distinct structures — the adrenal cortex and the adrenal medulla — that function as separate endocrine glands.

Adrenal cortex

The *adrenal cortex* is the large outer layer. It forms the bulk of the adrenal gland. It has three zones, or cell layers:
• *zona glomerulosa*, the outermost zone, which produces mineralocorticoids, primarily aldosterone
• *zona fasciculata*, the middle and largest zone, which produces the glucocorticoids cortisol (hydrocortisone), cortisone, and corticosterone as well as small amounts of the sex hormones androgen and estrogen
• *zona reticularis*, the innermost zone, which produces mainly glucocorticoids and some sex hormones.

Adrenal medulla

The *adrenal medulla*, or inner layer of the adrenal gland, functions as part of the sympathetic nervous system and produces two catecholamines. Because catecholamines play an important role in the autonomic nervous system (ANS), the adrenal medulla is considered a neuroendocrine structure.

It's fascinating, really. The zona fasciculata produces cortisol, cortisone, corticosterone, and even small amounts of androgen and estrogen.

Pancreas

The *pancreas*, a triangular organ, is nestled in the curve of the duodenum, stretching horizontally behind the stomach and extending to the spleen.

Endo and exo

The pancreas performs both endocrine and exocrine functions. *Acinar cells* make up most of the gland and regulate pancreatic exocrine function.

Islets of Langerhans

The endocrine cells of the pancreas are called the *islet cells*, or *islets of Langerhans.* These cells exist in clusters and are found scattered among the acinar cells. The islets contain alpha, beta, and delta cells that produce important hormones.

• Alpha cells produce *glucagon*, a hormone that raises the blood glucose level by triggering the breakdown of glycogen to glucose.

• Beta cells produce *insulin.* Insulin lowers the blood glucose level by stimulating the conversion of glucose to glycogen.

• Delta cells produce *somatostatin.* Somatostatin inhibits the release of GH, corticotropin, and certain other hormones.

I'm an alpha cell.

I'm a beta cell.

I guess I'm a delta cell.

Thymus

The *thymus* is located below the sternum and contains lymphatic tissue. It reaches maximal size at puberty and then starts to atrophy.

Producing Mr. T cells, fool!

Because the thymus produces T cells, which are important in cell-mediated immunity, its major role seems to be related to the immune system. However, the thymus also produces the peptide hormones thymosin and thymopoietin. These hormones promote growth of peripheral lymphoid tissue.

That's really neat. I didn't know I was made in the thymus.

Pineal gland

The tiny *pineal gland* lies at the back of the third ventricle of the brain. It produces the hormone *melatonin*, which may play a role in the neuroendocrine reproductive axis as well as other widespread actions.

Gonads

The *gonads* include the ovaries (in females) and the testes (in males).

Ovaries

The *ovaries* are paired, oval glands that are situated on either side of the uterus. They produce ova (eggs) and the steroidal hormones estrogen and progesterone. These hormones have four functions:

They promote development and maintenance of female sex characteristics.

They regulate the menstrual cycle.

They maintain the uterus for pregnancy.

Along with other hormones, they prepare the mammary glands for lactation.

Testes

The *testes* are paired structures that lie in an extra-abdominal pouch (scrotum) in the male. They produce spermatozoa and the male sex hormone testosterone. Testosterone stimulates and maintains male sex characteristics.

Hormones

Hormones are complex chemical substances that trigger or regulate the activity of an organ or a group of cells. Hormones are classified by their molecular structure as polypeptides, steroids, or amines.

Polypeptides

Polypeptides are protein compounds made of many amino acids that are connected by peptide bonds. They include:
• anterior pituitary hormones (GH, TSH, FSH, LH, and prolactin)
• posterior pituitary hormones (ADH and oxytocin)
• parathyroid hormone (PTH)
• pancreatic hormones (insulin and glucagon).

Steroids

Steroids are derived from cholesterol. They include:
• adrenocortical hormones secreted by the adrenal cortex (aldosterone and cortisol)
• sex hormones secreted by the gonads (estrogen and progesterone in females and testosterone in males).

Amines

Amines are derived from *tyrosine*, an essential amino acid found in most proteins. They include:
• thyroid hormones (T_4 and T_3)
• catecholamines (epinephrine, norepinephrine, and dopamine).

Hormone release and transport

Although all hormone release results from endocrine gland stimulation, release patterns of hormones vary greatly. For example:
• ACTH (secreted by the anterior pituitary) and cortisol (secreted by the adrenal cortex) are released in spurts in response to body rhythm cycles. Levels of these hormones peak in the morning.
• Secretion of PTH (by the parathyroid gland) and prolactin (by the anterior pituitary) occurs fairly evenly throughout the day.
• Secretion of insulin by the pancreas has both steady and sporadic release patterns.

Hormonal action

When a hormone reaches its target site, it binds to a specific receptor on the cell membrane or within the cell. Polypeptides and some amines bind to membrane receptor sites. The smaller, more lipid-soluble steroids and thyroid hormones diffuse through the cell membrane and bind to intracellular receptors.

Right on target!

After binding occurs, each hormone produces unique physiologic changes, depending on its target site and its specific action at that site. A particular hormone may have different effects at different target sites.

Thyroid and steroid hormones circulate while bound to plasma proteins, whereas catecholamines and most polypeptides aren't protein-bound.

Hormonal regulation

To maintain the body's delicate equilibrium, a feedback mechanism regulates hormone production and secretion. The mechanism involves hormones, blood chemicals and metabolites, and the nervous system. This system may be simple or complex. (See *The feedback loop*, page 110.)

Signaling secretory cells

For normal function, each gland must contain enough appropriately programmed secretory cells to release active hormone on demand.

Secretory cells need supervision. A secretory cell can't sense on its own when to release the hormone or how much to release. It gets this information from sensing and signaling systems that integrate many messages. Together, stimulatory and inhibitory signals actively control the rate and duration of hormone release.

When released, the hormone travels to *target cells*, where a receptor molecule recognizes it and binds to it. (See *Target cells*, page 111.)

Mechanisms that control hormone release

Four basic mechanisms control hormone release:

 the pituitary-target gland axis

 the hypothalamic-pituitary-target gland axis

Now I get it!

The feedback loop

This diagram shows the negative feedback mechanism that helps regulate the endocrine system.

From simple...
Simple feedback occurs when the level of one substance regulates the secretion of hormones (simple loop). For example, a low serum calcium level stimulates the parathyroid gland to release parathyroid hormone (PTH). PTH, in turn, promotes resorption of calcium. A high serum calcium level inhibits PTH secretion.

...to complex
When the hypothalamus receives negative feedback from target glands, the mechanism is more complicated (complex loop). *Complex feedback* occurs through an axis established between the hypothalamus, pituitary gland, and target organ. For example, secretion of corticotropin-releasing hormone from the hypothalamus stimulates release of corticotropin by the pituitary, which in turn stimulates cortisol secretion by the adrenal gland (the target organ). A rise in serum cortisol levels inhibits corticotropin secretion by decreasing corticotropin-releasing hormone.

Feedback refers to information sent to endocrine glands, signaling the need for changes in hormone levels.

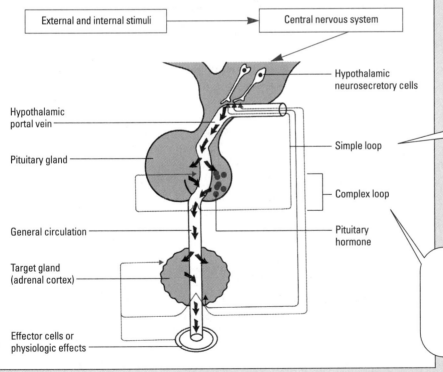

External and internal stimuli

Central nervous system

Hypothalamic neurosecretory cells

Hypothalamic portal vein

Simple loop

Pituitary gland

Complex loop

Pituitary hormone

General circulation

Target gland (adrenal cortex)

Effector cells or physiologic effects

Simple feedback occurs when the level of one substance regulates the secretion of hormones.

Complex feedback occurs through an axis established between the hypothalamus, pituitary gland, and target organ.

Now I get it!

Target cells

A hormone acts only on cells that have receptors specific to that hormone. The sensitivity of a target cell depends on how many receptors it has for a particular. The more receptor sites, the more sensitive the target cell.

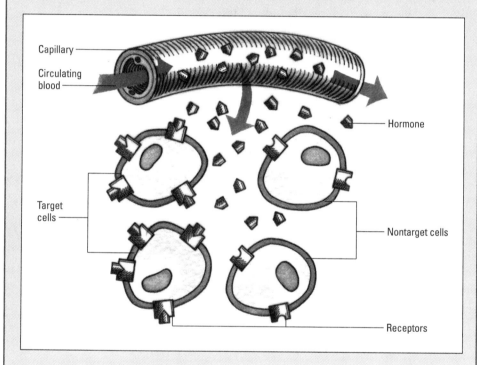

A hormone acts only on a cell that has a receptor specific to that hormone.

 chemical regulation

nervous system regulation.

Pituitary-target gland axis

The pituitary gland regulates other endocrine glands — and their hormones — through secretion of *trophic hormones* — releasing and inhibiting hormones. These hormones include:
• corticotropin, which regulates adrenocortical hormones

- TSH, which regulates T_4 and T_3
- LH, which regulates gonadal hormones.

Feedback

The pituitary gets feedback about target glands by continuously monitoring levels of hormones produced by these glands. If a change occurs, the pituitary corrects it in one of two ways:

- by increasing the trophic hormones, which stimulate the target gland to increase production of target gland hormones
- by decreasing the trophic hormones, thereby decreasing target gland stimulation and target gland hormone levels.

Hypothalamic-pituitary-target gland axis

The hypothalamus also produces trophic hormones that regulate anterior pituitary hormones. By controlling anterior pituitary hormones, which regulate the target gland hormones, the hypothalamus affects target glands as well.

Chemical regulation

Endocrine glands not controlled by the pituitary gland may be controlled by specific substances that trigger gland secretions. For example, blood glucose level is a major regulator of glucagon and insulin release. When the blood glucose level rises, the pancreas is stimulated to increase insulin secretion and suppress glucagon secretion. A depressed level of blood glucose, on the other hand, triggers increased glucagon secretion and suppresses insulin secretion.

Nervous system regulation

The central nervous system (CNS) helps to regulate hormone secretion in several ways.

Hypothalamus has control...

The hypothalamus controls pituitary hormones. Because hypothalamic nerve cells stimulate the posterior pituitary to produce ADH and oxytocin, these hormones are controlled directly by the CNS.

Specific substances, such as blood glucose, can trigger gland secretions.

...but stimuli matter, too

Nervous system stimuli — such as hypoxia (oxygen deficiency), nausea, pain, stress, and certain drugs — also affect ADH levels.

ANS steers this ship...

The ANS controls catecholamine secretion by the adrenal medulla.

...while stress spikes corticotropin

The nervous system also affects other endocrine hormones. For example, stress, which leads to sympathetic stimulation, causes the pituitary to release corticotropin.

Quick quiz

1. The purpose of the endocrine system is to:
   A. deliver nutrients to the body's cells.
   B. regulate and integrate the body's metabolic activities.
   C. eliminate waste products from the body.
Answer: B. Along with the nervous system, the endocrine system regulates and integrates the body's metabolic activities.

2. The mechanism that helps regulate the endocrine system is called the:
   A. transport mechanism.
   B. self-regulation mechanism.
   C. feedback mechanism.
Answer: C. The negative feedback mechanism helps regulate the endocrine system by signaling to the endocrine glands the need for changes in hormone levels.

3. The gland that produces glucagon is the:
   A. pancreas.
   B. thymus.
   C. adrenal gland.

Answer: A. The alpha cells of the pancreas produce glucagon, a hormone that raises the blood glucose level by triggering the breakdown of glycogen to glucose.

4.   Pituitary hormones are controlled by the:
     A.   pancreas.
     B.   hypothalamus.
     C.   thyroid gland.
Answer: B. The hypothalamus controls pituitary hormones.

Scoring

☆☆☆   If you answered all four questions correctly, astonishing!
        Your brain cells must be on steroids!

☆☆   If you answered two or three questions correctly — wow!
        You've just won a trip to the islets of Langerhans!
        Bon voyage!

☆   If you answered fewer than two questions correctly, don't
        moan over these hormones. There are 10
        more quick quizzes ahead.

Cardiovascular system

Just the facts

In this chapter, you'll learn:

♦ the structure of the heart

♦ how the heart contracts

♦ how the heart helps blood flow.

Introducing the cardiovascular system

The cardiovascular system (sometimes called the circulatory system) consists of the *heart*, *blood vessels*, and *lymphatics*. This network brings life-sustaining oxygen and nutrients to the body's cells, removes metabolic waste products, and carries hormones from one part of the body to another.

Doing double duty

The heart is actually two separate pumps: The right side pumps the blood to the lungs, and the left side pumps the blood to the rest of the body.

Where the heart lies

About the size of a closed fist, the heart lies beneath the sternum in the *mediastinum* (the cavity between the lungs), between the second and sixth ribs. In most people, the heart rests obliquely, with its right side below and almost in front of the left. Because of its oblique angle, the heart's broad part or top is at its upper right, and its pointed end (apex) is at its lower left. The apex is the

The angle at which I lie in the chest varies according to body build. In a tall, thin person, I'm more vertical than in a short, stocky person.

point of maximal impulse, where the heart sounds are the loudest.

Heart structure

Surrounded by a sac called the *pericardium*, the heart has a wall made up of three layers: *myocardium*, *endocardium*, and *epicardium*. Within the heart lie four chambers (two atria and two ventricles) and four valves (two atrioventricular [AV] and two semilunar valves). (See *Inside the heart*.)

Zoom in

Inside the heart

Within the heart lie four chambers (two atria and two ventricles) and four valves (two atrioventricular and two semilunar valves). A system of blood vessels carries blood to and from the heart.

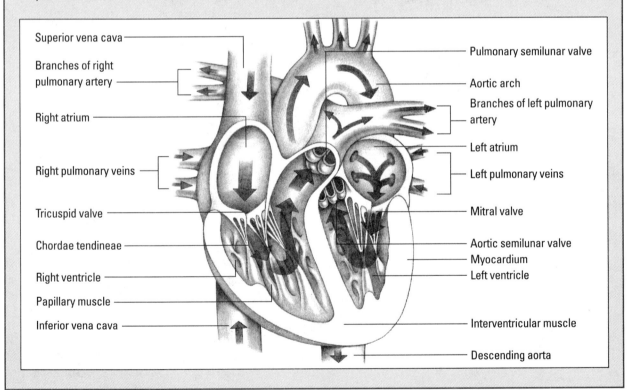

Pericardium

The *pericardium* is a fibroserous sac that surrounds the heart and the roots of the *great vessels* (those vessels that enter and leave the heart). It consists of the fibrous pericardium and the serous pericardium.

Fibrous fits freely

The *fibrous pericardium*, composed of tough, white fibrous tissue, fits loosely around the heart, protecting it.

Let's get serous

The *serous pericardium*, the thin, smooth inner portion, has two layers:
• The *parietal layer* lines the inside of the fibrous pericardium.
• The *visceral layer* adheres to the surface of the heart.

Slippin' and slidin'

Between the fibrous and serous pericardium is the *pericardial space. Pericardial fluid* lubricates the surfaces of the space and allows the heart to move easily during contraction.

I'm supported and protected by a tough, fibrous sac but I stay comfortable because it has a smooth inner lining.

The wall

The wall of the heart consists of three layers:
• The *epicardium*, the outer layer (and the visceral layer of the serous pericardium), is made up of squamous epithelial cells overlying connective tissue.
• The *myocardium*, the middle layer, forms most of the heart wall. It has striated muscle fibers that cause the heart to contract.
• The *endocardium*, the heart's inner layer, consists of endothelial tissue with small blood vessels and bundles of smooth muscle.

The chambers

The heart contains four hollow chambers: two atria (singular: atrium) and two ventricles.

Upstairs...

The *atria*, the upper chambers, are separated by the *interatrial septum*. They receive blood returning to the heart and pump blood to the ventricles.

...where the blood comes in

The *right atrium* receives blood from the *superior* and *inferior venae cavae*. The *left atrium*, which is smaller but has thicker walls than the right atrium, forms the uppermost part of the heart's left border. It receives blood from the two pulmonary veins.

Downstairs...

The *right* and *left ventricles*, separated by the *interventricular septum*, make up the two lower chambers. The ventricles receive blood from the atria. Composed of highly developed musculature, the ventricles are larger and thicker walled than the atria.

At rest, only about 20% of my blood goes to skeletal muscles. When I exercise, that percentage can increase to 70%.

...where the blood goes out

The right ventricle pumps blood to the lungs. The left ventricle, larger than the right, pumps blood through all other vessels of the body.

The valves

The heart contains four valves, two *AV valves* and two *semilunar valves*.

One way only

The valves allow forward flow of blood through the heart and prevent backward flow. They open and close in response to pressure changes caused by ventricular contraction and blood ejection. The two AV valves separate the atria from the ventricles. One of the two semilunar valves is the *pulmonic valve*, which prevents backflow from the pulmonary artery into the right ventricle. The other semilunar valve is the *aortic valve*, which prevents backflow from the aorta into the left ventricle.

On the cusps

The right AV valve, also called the *tricuspid valve*, has three triangular *cusps*, or leaflets. The left AV valve,

Memory jogger

If you can remember that there are two distinct heart sounds, you can recall that there are two sets of heart valves. Closure of the atrioventricular valves makes the first heart sound, the *Lub*; closure of the semilunar valves makes the second heart sound, the *Dub*.

called the *mitral* or *bicuspid valve*, contains two cusps, a large anterior and a smaller posterior. *Chordae tendineae* attach the cusps of the AV valves to papillary muscles in the ventricles. The semilunar valves have three cusps that are shaped like half-moons.

The heart's contraction

Contraction of the heart, occurring as a result of its *conduction system*, causes the blood to move throughout the body. (See *Cardiac conduction system.*)

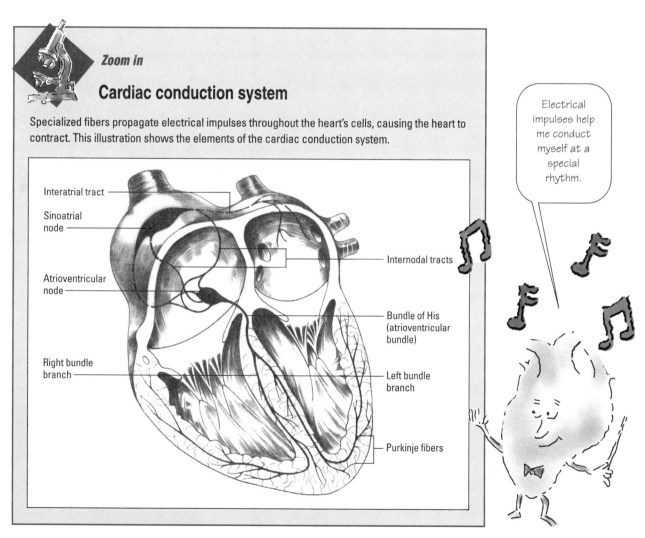

Zoom in

Cardiac conduction system

Specialized fibers propagate electrical impulses throughout the heart's cells, causing the heart to contract. This illustration shows the elements of the cardiac conduction system.

Interatrial tract

Sinoatrial node

Atrioventricular node

Right bundle branch

Internodal tracts

Bundle of His (atrioventricular bundle)

Left bundle branch

Purkinje fibers

Electrical impulses help me conduct myself at a special rhythm.

Setting the pace

The conduction system of the heart contains *pacemaker cells*, which have three unique characteristics:
- *automaticity*, the ability to generate an electrical impulse automatically
- *conductivity*, the ability to pass the impulse to the next cell
- *contractility*, the ability to shorten the fibers in the heart when receiving the impulse.

Specialized pacemaker cells in my myocardium allow electrical impulse conduction.

Feeling impulsive

The *sinoatrial (SA) node*, located on the *endocardial surface* of the right atrium, near the superior vena cava, is the normal pacemaker of the heart, generating an impulse between 60 and 100 times per minute. The SA node's firing spreads an impulse throughout the right and left atria, resulting in *atrial contraction*.

Fill 'er up

The *AV node*, situated low in the *septal wall* of the right atrium, slows the impulse conduction between the atria and ventricles. This "resistor" node allows time for the contracting atria to fill the ventricles with blood before the lower chambers contract.

Spreading the word — "contract"

From the AV node, the impulse travels to the *bundle of His* (modified muscle fibers), branching off to the right and left bundles. Finally, the impulse travels to the *Purkinje fibers*, the distal portions of the left and right bundle branches. These fibers fan across the surface of the ventricles from the endocardium the myocardium. As the impulse spreads, it brings "the word" to the blood-filled ventricles to contract.

My conduction system has two backup impulse generators.

Foolproof

The conduction system has two built-in safety mechanisms. If the SA node fails to fire, the AV node will generate an impulse between 40 and 60 times per minute. If the SA node and AV node fail, the ventricles can generate their own impulse between 20 and 40 times per minute.

Cardiac cycle

The *cardiac cycle* is the period from the beginning of one heartbeat to the beginning of the next. During this cycle, electrical and mechanical events must occur in the proper sequence and to the proper degree to provide adequate blood flow to all body parts. The cardiac cycle has two phases — *systole* and *diastole*. (See *Phases of the cardiac cycle*, page 122.)

Contract...

At the beginning of *systole*, the ventricles contract. Increasing blood pressure in the ventricles forces the AV valves (mitral and tricuspid) to close and the semilunar valves (pulmonic and aortic) to open.

As the ventricles contract, ventricular blood pressure builds until it exceeds the pressure in the pulmonary artery and the aorta. Then the semilunar valves open, and the ventricles eject blood into the aorta and the pulmonary artery.

...and release

When the ventricles empty and relax, ventricular pressure falls below that in the pulmonary artery and the aorta. At the beginning of *diastole*, the semilunar valves close to prevent the backflow of blood into the ventricles, and the mitral and tricuspid valves open, allowing blood to flow into the ventricles from the atria.

When the ventricles become full, near the end of this phase, the atria contract to send the remaining blood to the ventricles. Then a new cardiac cycle begins as the heart enters systole again.

The specs for this cycle's engine

Cardiac output refers to the amount of blood the heart pumps in 1 minute. It's equal to the heart rate multiplied by the *stroke volume*, the amount of blood ejected with each heartbeat. Stroke volume, in turn, depends on three major factors: *contractility*, *preload*, and *afterload*. (See *Understanding preload, afterload, and contractility*, page 123.)

The autonomic nervous system increases or decreases my heart activity to meet the metabolic needs of my body. Whew!

Now I get it!

Phases of the cardiac cycle

The cardiac cycle consists of the following five phases.

Isovolumetric ventricular contraction — In response to ventricular depolarization, tension in the ventricles increases. This rise in pressure within the ventricles leads to closure of the mitral and tricuspid valves. The pulmonic and aortic valves stay closed during the entire phase.

Ventricular ejection — When ventricular pressure exceeds aortic and pulmonary arterial pressure, the aortic and pulmonic valves open and the ventricles eject blood.

Isovolumetric relaxation — When ventricular pressure falls below the pressure in the aorta and pulmonary artery, the aortic and pulmonic valves close. All valves are closed during this phase. Atrial diastole occurs as blood fills the atria.

Atrial systole — Known as the atrial kick, atrial systole (coinciding with late ventricular diastole) supplies the ventricles with the remaining 30% of the blood for each heartbeat.

Ventricular filling — Atrial pressure exceeds ventricular pressure, which causes the mitral and tricuspid valves to open. Blood then flows passively into the ventricles. About 70% of ventricular filling takes place during this phase.

Now I get it!

Understanding preload, afterload, and contractility

If you think of the heart as a balloon, it will help you understand stroke volume.

Blowing up the balloon
Preload is the stretching of muscle fibers in the ventricles. This stretching results from blood volume in the ventricles at end-diastole. According to *Starling's law*, the more the heart muscles stretch during diastole, the more forcefully they contract during systole. Think of preload as the balloon stretching as air is blown into it. The more air the greater the stretch.

The balloon's stretch
Contractility refers to the inherent ability of the myocardium to contract normally. Contractility is influenced by preload. The greater the stretch the more forceful the contraction—or, the more air in the balloon, the greater the stretch, and the farther the balloon will fly when air is allowed to expel.

The knot that ties the balloon
Afterload refers to the pressure that the ventricular muscles must generate to overcome the higher pressure in the aorta to get the blood out of the heart. *Resistance* is the knot on the end of the balloon, which the balloon has to work against to get the air out.

The heart and blood flow

As blood makes its way through the vascular system, it travels through five distinct types of blood vessels:
- arteries
- arterioles
- capillaries
- venules
- veins.

Vessel structure

The structure of each type of vessel differs according to its function in the cardiovascular system and the pressure exerted by the volume of blood at various sites within the system.

Through thick...

Arteries have thick, muscular walls to accommodate the flow of blood at high speeds and pressures. *Arterioles* have thinner walls than arteries. They constrict or dilate to control blood flow to the *capillaries*, which (being microscopic) have walls composed of only a single layer of endothelial cells.

...and thin

Venules gather blood from the capillaries; their walls are thinner than those of arterioles. *Veins* have thinner walls than arteries but have larger diameters because of the low blood pressures of venous return to the heart.

Taking the long way home

About 60,000 miles of arteries, arterioles, capillaries, venules, and veins keep blood circulating to and from every functioning cell in the body. (See *Major blood vessels.*)

Circulation

There are three methods of circulation that carry blood throughout the body: *pulmonary, systemic,* and *coronary.*

Pulmonary circulation

Blood travels to the lungs to pick up oxygen and release carbon dioxide.

Picking it up...

As the blood moves from the heart, to the lungs, and back again, it proceeds as follows:
• Unoxygenated blood travels from the right ventricle through the pulmonic valve into the *pulmonary arteries.*
• Blood passes through progressively smaller arteries and arterioles into the capillaries of the lungs.

Body shop

Major blood vessels

This illustration shows the body's major arteries and veins.

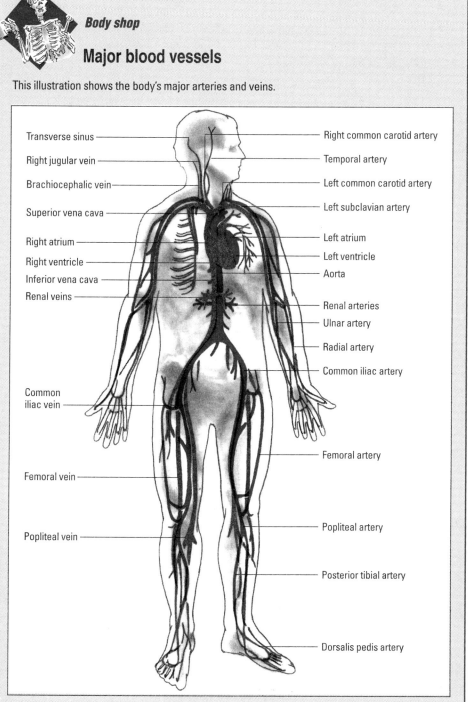

Transverse sinus

Right jugular vein

Brachiocephalic vein

Superior vena cava

Right atrium

Right ventricle

Inferior vena cava

Renal veins

Common iliac vein

Femoral vein

Popliteal vein

Right common carotid artery

Temporal artery

Left common carotid artery

Left subclavian artery

Left atrium

Left ventricle

Aorta

Renal arteries

Ulnar artery

Radial artery

Common iliac artery

Femoral artery

Popliteal artery

Posterior tibial artery

Dorsalis pedis artery

Add the arterioles, venules, and capillaries and you've got nearly 60,000 miles of blood vessels.

• Blood reaches the *alveoli* and exchanges carbon dioxide for oxygen.
• Oxygenated blood then returns via venules and veins to the *pulmonary veins*, which carry it back to the heart's left atrium.

Systemic circulation

Blood pumped from the left ventricle carries oxygen and other nutrients to body cells and transports waste products for excretion.

...carrying it out

The major artery, the *aorta*, branches into vessels that supply specific organs and areas of the body. As it arches out of the top of the heart and down to the abdomen, three arteries branch off the top of the arch to supply the upper body with blood:
• The *left common carotid artery* supplies blood to the brain.
• The *left subclavian artery* supplies the arms.
• The *innominate artery* supplies the upper chest.
   As the aorta descends through the thorax and abdomen, its branches supply the organs of the GI and genitourinary systems, spinal column, and lower chest and abdominal muscles. Then the aorta divides into the *iliac arteries*, which further divide into *femoral arteries*.

Division = addition = perfusion

As the arteries divide into smaller units, the number of vessels increases dramatically, thereby increasing the area of tissue to which blood flows, also called the *area of perfusion*.

Dilation is another part of the equation

At the end of the arterioles and the beginning of the capillaries, strong *sphincters* control blood flow into the tissues. These sphincters dilate to permit more flow when needed, close to shunt blood to other areas, or constrict to increase blood pressure.

A large area of low pressure

Although the *capillary bed* contains the smallest vessels, it supplies blood to the largest number of cells. Capillary pressure is extremely low to allow for the exchange of

My pumping action can be felt at certain sites. This regular expansion and contraction of arteries is called the pulse.

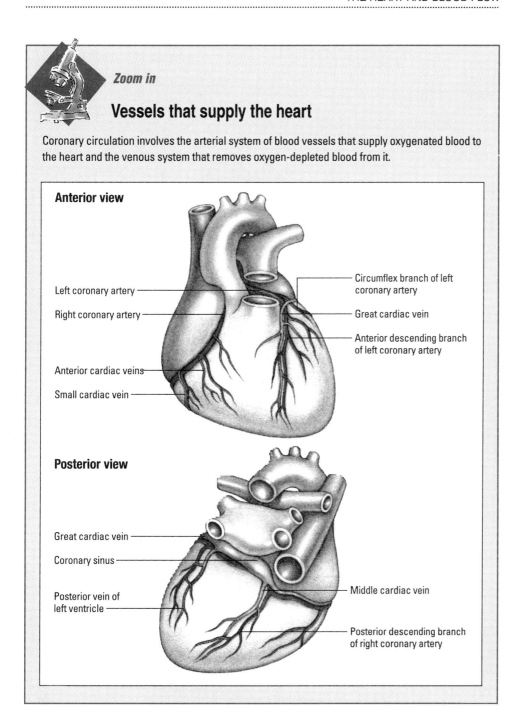

Zoom in

Vessels that supply the heart

Coronary circulation involves the arterial system of blood vessels that supply oxygenated blood to the heart and the venous system that removes oxygen-depleted blood from it.

Anterior view

Left coronary artery

Right coronary artery

Anterior cardiac veins

Small cardiac vein

Circumflex branch of left coronary artery

Great cardiac vein

Anterior descending branch of left coronary artery

Posterior view

Great cardiac vein

Coronary sinus

Posterior vein of left ventricle

Middle cardiac vein

Posterior descending branch of right coronary artery

nutrients, oxygen, and carbon dioxide with body cells. From the capillaries, blood flows into venules and, eventually, into veins.

No backflow

Valves in the veins prevent blood backflow. Pooled blood in each valved segment is moved toward the heart by pressure from the moving volume of blood from below. The veins merge until they form two main branches, the *superior vena cava* and *inferior vena cava*, that return blood to the right atrium.

Coronary circulation

The heart relies on the coronary arteries and their branches for its supply of oxygenated blood and depends on the cardiac veins to remove oxygen-depleted blood. (See *Vessels that supply the heart*, page 127.)

Back home, dishing it out

During left ventricular systole, blood is ejected into the aorta. During diastole, blood flows out of the heart and then through the coronary arteries to nourish the heart muscle.

From the right...

The *right coronary artery* supplies blood to the right atrium, part of the left atrium, most of the right ventricle, and the inferior part of the left ventricle.

...and from the left

The *left coronary artery,* which splits into the *anterior descending artery* and *circumflex artery,* supplies blood to the left atrium, most of the left ventricle, and most of the interventricular septum.

Superficially speaking

The *cardiac veins* lie super-ficial to the arteries. The largest vein, the *coronary sinus,* opens into the right atrium. Most of the major cardiac veins empty into the coronary sinus, except for the *anterior cardiac veins,* which empty into the right atrium.

Like any other muscle, I also need oxygen to function. Ah, there's nothing like a little fresh air!

Quick quiz

1. During systole the ventricles contract. This causes:
   A.   all four heart valves to close.
   B.   the AV valves to close and the semilunar valves to open.
   C.   the AV valves to open and the semilunar valves to close.

Answer: B. During systole the pressure is greater in the ventricles than in the atria, causing the AV valves (the tricuspid and mitral valves) to close. The pressure in the ventricles is also greater than the pressure in the aorta and pulmonary artery, forcing the semilunar valves (the pulmonic and aortic valves) to open.

2. The normal pacemaker of the heart is:
   A.   the SA node.
   B.   the AV node.
   C.   the ventricles.

Answer: A. The SA node is the normal pacemaker of the heart, generating impulses 60 to 100 times per minute. The AV node is the secondary pacemaker of the heart (40 to 60 beats per minute). The ventricles are the last line of defense (20 to 40 beats per minute).

3. The pressure the ventricular muscle must generate to overcome the higher pressure in the aorta refers to:
   A.   contractility.
   B.   preload.
   C.   afterload.

Answer: C. Afterload is the pressure the ventricular muscle must generate to overcome the higher pressure in the aorta to get the blood out of the heart.

4. The vessels that carry oxygenated blood back to the heart and left atrium are:
   A.   capillaries.
   B.   pulmonary veins.
   C.   pulmonary arteries.

Answer: B. Oxygenated blood returns by way of venules and veins to the pulmonary veins, which carry it back to the heart's left atrium.

5. The layer of the heart responsible for contraction is:
   A. myocardium.
   B. pericardium.
   C. endocardium.

Answer: A. The myocardium has striated muscle fibers that cause the heart to contract.

Scoring

☆☆☆   If you've answered all five questions correctly, marvelous! You've gotten to the heart of the cardiovascular system.

☆☆   If you've answered three or four questions correctly, great! We won't say you're "vein" if you're a little proud of yourself.

☆   If you've answered fewer than three questions correctly, take heart! It's time to circulate on to the next chapter.

Hematologic system

Just the facts

In this chapter, you'll learn:

♦ how blood cells develop

♦ the functions of the different blood components

♦ how blood cells clot

♦ blood groups and their significance.

Introducing the hematologic system

The hematologic system consists of the blood and bone marrow. Blood delivers oxygen and nutrients to all tissues, removes wastes, and transports gases, blood cells, immune cells, and hormones throughout the body.

Living up to their potential

The hematologic system manufactures new blood cells through a process called *hematopoiesis. Multipotential stem cells* in bone marrow give rise to five distinct cell types, called *unipotential stem cells.* Unipotential cells differentiate into one of the following types of blood cells:

• erythrocyte
• granulocyte
• agranulocyte
• platelet. (See *Tracing blood cell formation*, pages 132 and 133.)

(Text continues on page 134.)

Now I get it!

Tracing blood cell formation

Blood cells form and develop in the bone marrow by a process called *hematopoiesis*. This chart breaks down the process from when the five unipotential stem cells are "born" from the multipotential stem cell until they each reach "adulthood" as fully formed cells—either erythrocytes, granulocytes, agranulocytes, or platelets.

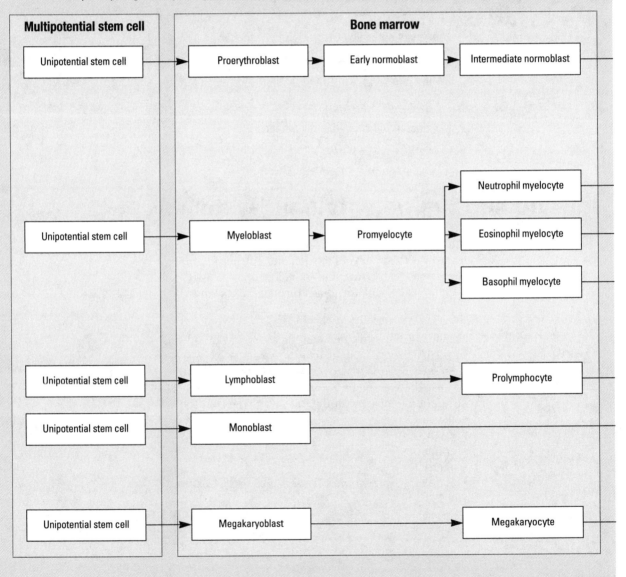

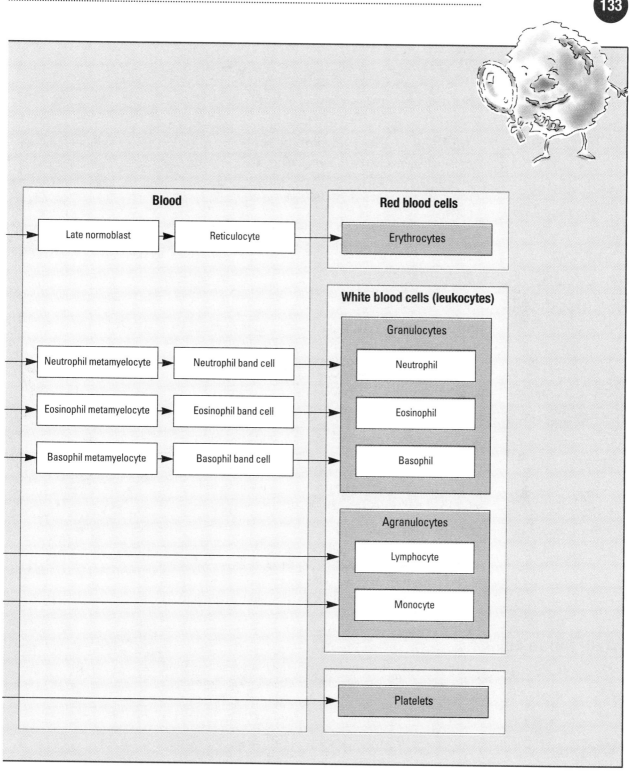

Blood

Late normoblast → Reticulocyte

Neutrophil metamyelocyte → Neutrophil band cell

Eosinophil metamyelocyte → Eosinophil band cell

Basophil metamyelocyte → Basophil band cell

Red blood cells

Erythrocytes

White blood cells (leukocytes)

Granulocytes

Neutrophil

Eosinophil

Basophil

Agranulocytes

Lymphocyte

Monocyte

Platelets

Blood components

Blood consists of various formed elements, or *blood cells*, suspended in a fluid called *plasma*.

The RBCs of blood — and the WBCs and platelets, too

Formed elements in the blood include:
- *red blood cells (RBCs)*, or *erythrocytes*
- *white blood cells (WBCs)*, or *leukocytes*
- platelets.

RBCs and platelets function entirely within blood vessels; WBCs act mainly in the tissues outside the blood vessels.

Red blood cells

RBCs transport oxygen and carbon dioxide to and from body tissues. They contain *hemoglobin*, the oxygen-carrying substance that gives blood its red color. The RBC surface carries *antigens*, which determine a person's *blood group*, or blood type.

The life and times of the RBC

RBCs have an average life span of 120 days. Bone marrow releases RBCs into circulation in immature form as *reticulocytes*. The reticulocytes mature into RBCs in about 1 day. The spleen sequesters, or isolates, old, worn-out RBCs, removing them from circulation.

A balance between removal and renewal

The rate of reticulocyte release usually equals the rate of old RBC removal. When RBC depletion occurs (for example, with hemorrhage) the bone marrow increases reticulocyte production to maintain the normal RBC count.

The body manufactures billions of new RBCs like me every day!

White blood cells

Five types of WBCs participate in the body's defense and immune systems. These cells are classified as *granulocytes* or *agranulocytes*.

Granulocytes

Granulocytes include *neutrophils, eosinophils,* and *basophils* — collectively known as *polymorphonuclear leukocytes.* All granulocytes contain a single *multilobular nucleus* and *granules* in the cytoplasm. Each cell type exhibits different properties and each is activated by different stimuli.

Swallowing up your enemies

Neutrophils, the most numerous granulocytes, account for 47.6% to 76.8% of circulating WBCs. These phagocytic cells engulf, ingest, and digest foreign materials. They leave the bloodstream by passing through the capillary walls into the tissues (a process called *diapedesis),* then migrate to and accumulate at infection sites.

Yikes! A neutrophil is after us!

I eat scum like you for breakfast!

The fatigued and their replacements

Worn-out neutrophils form the main component of pus. Bone marrow produces their replacements, immature neutrophils called *bands.* In response to infection, bone marrow must produce many immature cells and release them into circulation, elevating the band count.

Allies against allergies

Eosinophils account for 0.3% to 7% of circulating WBCs. These granulocytes also migrate from the bloodstream by diapedesis but do so as a response to an allergic reaction. Eosinophils accumulate in loose connective tissue, where they become involved in the ingestion of antigen-antibody complexes.

Fighting the flames

Basophils usually constitute fewer than 2% of circulating WBCs. They possess little or no phagocytic ability. Their cytoplasmic granules secrete *histamine* in response to certain inflammatory and immune stimuli. Histamine makes the blood vessels more permeable and eases the passage of fluids from the capillaries into body tissues.

Memory jogger

Two other "N" words, numerous and neutralize, should help you remember what neutrophils do.

Agranulocytes

WBCs in the agranulocyte category — *monocytes* and *lymphocytes* — lack specific cytoplasmic granules and have nuclei without lobes. (See *Comparing granulocytes and agranulocytes.*)

The few and the large

Monocytes, the largest of the WBCs, constitute only 0.6% to 9.6% of WBCs in circulation. Like neutrophils, monocytes are phagocytic and enter the tissues by diapedesis. Outside the bloodstream, monocytes enlarge and mature, becoming tissue *macrophages* (also called *histiocytes*).

Protection against infection

As macrophages, monocytes may roam freely through the body when stimulated by inflammation. Usually, they remain immobile, populating most organs and tissues. Collectively, they serve as components of the *reticuloendothelial system,* which defends the body against infection and disposes of cell breakdown products.

Fluid finders

Macrophages concentrate in structures that filter large amounts of body fluid, such as the liver, spleen, and lymph nodes, where they defend against invading organisms. Macrophages are efficient *phagocytes,* cells that ingest microorganisms, cellular debris (including worn-out neutrophils), and necrotic tissue. When mobilized at an infection site, they phagocytize cellular remnants and promote wound healing.

Last and, in fact, least (in size)

Lymphocytes, the smallest of the WBCs and the second most numerous (16.2% to 43%), derive from stem cells in the bone marrow. There are two types of lymphocytes:
• *T lymphocytes* directly attack an infected cell.
• *B lymphocytes* produce antibodies against specific antigens.

We granulocytes serve as the body's first line of cellular defense.

Platelets

Platelets are small, colorless, disk-shaped cytoplasmic fragments split from cells in bone marrow called *megakaryocytes.*

Now I get it!

Comparing granulocytes and agranulocytes

White blood cells (WBCs) are like soldiers fighting off the enemy. Each type of WBC fights a different enemy.

On the front line

Granulocytes, with "platoons" of basophils, neutrophils, and eosinophils, are the first forces "marshalled" against invading foreign organisms.

Basophils spit histamine at inflammatory and immune stimuli.	Neutrophils eat foreign bodies.	Eosinophils eat antigens and antibodies.
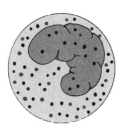	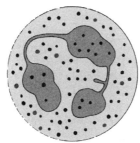	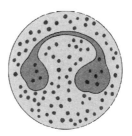

In the trenches

Agranulocytes, with "platoons" of lymphocytes and monocytes, may roam freely on "patrol" when inflammation is reported, but they mainly "dig in" at structures that filter large amounts of fluid (such as the liver) and defend against invaders.

Lymphocytes eat the enemy.	Monocytes eat bacteria, cellular debris, and necrotic tissue.
	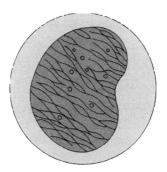

On the plate(lets)

These fragments, which have a life span of approximately 10 days, perform three vital functions:
- initiating contraction of damaged blood vessels to minimize blood loss
- forming *hemostatic plugs* in injured blood vessels
- with plasma, providing materials that accelerate blood coagulation.

When a blood vessel is injured or severed, clotting begins within minutes.

How blood cells clot

Hemostasis is the complex process by which *platelets*, *plasma*, and *coagulation factors* interact to control bleeding.

Stop the bleeding!

When a blood vessel ruptures, local *vasoconstriction* (decrease in the caliber of blood vessels) and *platelet clumping* (aggregation) at the site of the injury initially help prevent hemorrhage. This activation of the coagulation system, called the *extrinsic cascade*, requires the release of *tissue thromboplastin* from the damaged cells.

A more long-term solution

However, formation of a more stable clot requires initiation of the complex clotting mechanisms known as the *intrinsic cascade system*. This clotting system is activated by a protein, called factor XII, one of 13 substances that are necessary for coagulation and derived from plasma and tissue.

Come together

The final result of the intrinsic cascade is a *fibrin clot*, an accumulation of a fibrous, insoluble protein at the site of the injury. (See *How blood clots.*)

Coagulation factors

The materials that platelets and plasma provide work with *coagulation factors* to serve as *precursor compounds* in the clotting (coagulation) of blood.

Now I get it!

How blood clots

When a blood vessel is severed or injured, three interrelated processes take place.

Constriction and aggregation

Immediately, the vessels affected by the injury contract *(constriction)*, reducing blood flow. Also, platelets, stimulated by the exposed collagen of the damaged cells, begin to clump together *(aggregation)*. Aggregation provides a temporary seal and a site for clotting to take place. The platelets release a number of substances that enhance constriction and aggregation.

Clotting pathways

Clotting, or *coagulation*—the transformation of blood from a liquid to a solid—may be initiated through two different pathways, the intrinsic pathway or the extrinsic pathway. The *intrinsic pathway* is activated when plasma comes in contact with damaged vessel surfaces. The *extrinsic pathway* is activated when tissue thromboplastin, a substance released by damaged endothelial cells, comes into contact with one of the clotting factors.

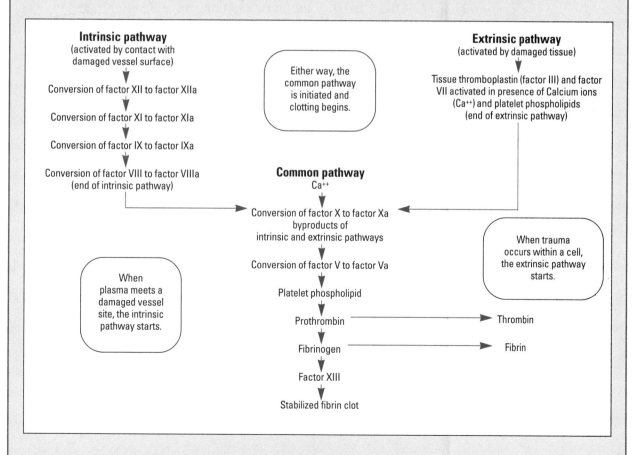

A baker's dozen

Designated by name and Roman numeral, these *coagulation factors* are activated in a chain reaction, each one in turn activating the next factor in the chain:

• Factor I, *fibrinogen*, is a high-molecular-weight protein synthesized in the liver and converted to fibrin during the coagulation cascade.

• Factor II, *prothrombin*, is a protein synthesized in the liver in the presence of vitamin K and converted to thrombin during coagulation.

• Factor III, *tissue thromboplastin*, is released from damaged tissue; it's required to initiate the second phase, the extrinsic cascade system.

• Factor IV, consisting of *calcium ions*, is required throughout the entire clotting sequence.

• Factor V, or *labile factor (proaccelerin)*, is a protein that is synthesized in the liver and functions during the combined pathway phase of the coagulation system.

• Factor VII, *serum prothrombin conversion accelerator* or *stable factor (proconvertin)*, is a protein synthesized in the liver in the presence of vitamin K; it's activated by Factor III in the extrinsic system.

• Factor VIII, *antihemophilic factor (antihemophilic globulin)*, is a protein synthesized in the liver and required during the instrinsic phase of the coagulation system.

• Factor IX, *plasma thromboplastin component*, a protein synthesized in the liver in the presence of vitamin K, is required in the intrinsic phase of the coagulation system.

• Factor X, *Stuart factor (Stuart-Prower factor)*, is a protein synthesized in the liver in the presence of vitamin K; it's required in the combined pathway of the coagulation system.

• Factor XI, *plasma thromboplastin antecedent*, is a protein synthesized in the liver and required in the intrinsic system.

• Factor XII, *Hageman factor*, is a protein required in the intrinsic system.

• Factor XIII, *fibrin stabilizing factor*, is a protein required to stabilize the fibrin strands in the combined pathway phase of the coagulation system.

When I'm damaged, I release thromboplastin, which activates the extrinsic portion of the coagulation system.

Coagulation factors stimulate clot formation without hindering blood flow.

Blood groups

Blood groups are determined by the presence or absence of genetically determined *antigens* or *agglutinogens* (glycoproteins) on the surface of RBCs. A, B, and Rh are the most clinically significant blood antigens.

ABO groups

Testing for the presence of A and B antigens on RBCs is the most important system for classifying blood:
- Type A blood has A antigen on its surface.
- Type B blood has B antigen.
- Type AB blood has both A and B antigens.
- Type O blood has neither A nor B antigen.

Opposites don't attract

Plasma may contain *antibodies* that interact with these antigens, causing the cells to *agglutinate*, or combine into a mass. However, plasma can't contain antibodies to its own cell antigen or it would destroy itself. Thus, type A blood has A antigen but no A antibodies; however, it does have B antibodies.

Making a match

Precise blood-typing and crossmatching (mixing and observing for agglutination of donor cells) are essential, especially for blood transfusions. A donor's blood must be compatible with a recipient's or the result can be fatal. The following blood groups are compatible:
- type A with type A or O
- type B with type B or O
- type AB with type A, B, AB, or O
- type O with type O only. (See *Reviewing blood type compatibility*, page 142.)

Rh typing

Rh typing determines whether the Rh factor is present or absent in the blood. Of the eight types of Rh antigens, only C, D, and E are common.

Now I get it!

Reviewing blood type compatibility

Precise blood typing and crossmatching can prevent the transfusion of incompatible blood, which can be fatal. Usually, typing the recipient's blood and crossmatching it with available donor blood take less than 1 hour.

Making a match

Agglutinogen (an antigen in red blood cells) and *agglutinin* (an antibody in plasma) distinguish the four ABO blood groups. This chart shows ABO compatibility from the perspectives of the recipient and the donor.

Blood group	Antibodies present in plasma	Compatible RBCs	Compatible plasma
Recipient			
O	Anti-A and anti-B	O	O, A, B, AB
A	Anti-B	A, O	A, AB
B	Anti-A	B, O	B, AB
AB	Neither anti-A nor anti-B	AB, A, B, O	AB
Donor			
O	Anti-A and anti-B	O, A, B, AB	O
A	Anti-B	A, AB	A, O
B	Anti-A	B, AB	B, O
AB	Neither anti-A nor anti-B	AB	AB, A, B, O

Memory jogger

Blood types are easy to remember because they're named after the antigens they contain—A or B or both A and B—except for type O, which contains neither. The O serves as a nice visual reminder of that absence.

Positive and negative types

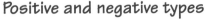

Typically, blood contains the Rh antigen. Blood with the Rh antigen is Rh-positive; blood without the Rh antigen is Rh-negative. Anti-Rh antibodies can appear only in a person who has become sensitized. Anti-Rh antibodies can appear in the blood of an Rh-negative person after entry of Rh-positive RBCs in the bloodstream—for example, from transfusion of Rh-positive blood. An Rh-negative female who carries an Rh-positive fetus may also acquire anti-Rh antibodies.

Quick quiz

1. The component of blood that triggers defense and immune responses is the:

    A.   WBC.

    B.   platelet.

    C.   RBC.

Answer: A. Because of their phagocytic capabilities, WBCs serve as the body's first line of cellular defense against foreign organisms.

2. The complex process by which platelets, plasma, and coagulation factors interact to control bleeding is called:

    A.   phagocytosis.

    B.   hematopoiesis.

    C.   hemostasis.

Answer: C. Hemostasis is achieved through a three-part process: vasoconstriction, platelet aggregation, and coagulation.

3. Blood cells form and develop in the:

    A.   platelet.

    B.   liver.

    C.   bone marrow.

Answer: C. Multipotential stem cells in the bone marrow give rise to five distinct cell types called unipotential stem cells. Each of these stem cells can differentiate into an erythrocyte, a granulocyte, an agranulocyte, or a platelet.

4. Blood groups are determined by testing for A and B antigens on the:

    A.   RBC.

    B.   WBC.

    C.   leukocyte.

Answer: A. Blood groups are determined by the presence or absence of antigens or agglutinogens on the surface of RBCs.

Scoring

☆☆☆   If you answered all four questions correctly, wonderful! You're clearly thinking hemato-logically!

☆☆   If you answered two or three questions correctly, great. You've coagulated all the information in this chapter into a solid understanding of blood.

☆   If you answered fewer than two questions correctly, be positive (or A positive, or AB positive). Just read the chapter again and give the quiz another chance.

Immune system

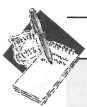

Just the facts

In this chapter, you'll learn:

♦ what organs and tissues make up the immune system

♦ how the immune system works

♦ what happens when the immune system fails.

Introducing the immune system

The immune system defends the body against invasion by harmful organisms or chemical toxins. The blood is an important part of this protective system.

A close relative

Although the immune system and blood are distinct entities, they're closely related. Their cells share a common origin in the bone marrow, and the immune system uses the bloodstream to transport its "troops" to the site of an invasion.

The major components

Organs and tissues of the immune system are referred to as "lymphoid" because they're all involved with the growth, development, and dissemination of lymphocytes, one type of white blood cells (WBCs). The immune system has three major divisions:
• central lymphoid organs and tissues
• peripheral lymphoid organs and tissues

Body shop

Organs and tissues of the immune system

The immune system includes organs and tissues in which lymphocytes predominate as well as cells that circulate in the blood. This illustration shows central, peripheral, and accessory lymphoid organs and tissue.

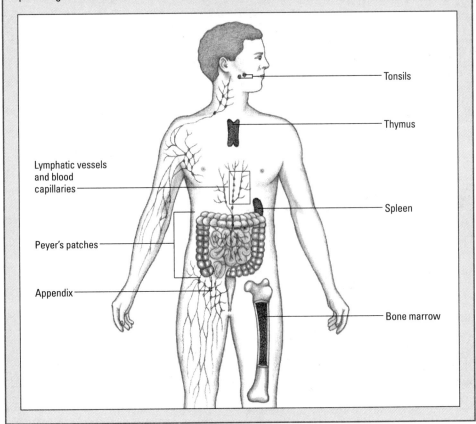

Tonsils

Thymus

Lymphatic vessels and blood capillaries

Spleen

Peyer's patches

Appendix

Bone marrow

• accessory lymphoid organs and tissues. (See *Organs and tissues of the immune system*.)

Central lymphoid organs and tissues

The bone marrow and thymus play a role in the development of B cells and T cells—the two major types of *lymphocytes.*

We're T and B cells, the two major types of lymphocytes.

Bone marrow

The *bone marrow* contains stem cells, which may develop into any of several different cell types. Such cells are *multipotential*, meaning they're capable of taking many forms. The immune system and blood cells develop from stem cells in a process called *hematopoiesis*.

To be B or to be T, that is the question...

Soon after their differentiation from other stem cells, some of the cells destined to become immune system cells serve as sources for *lymphocytes;* other cells of this differentiated group develop into *phagocytes* (cells that ingest microorganisms). Those that become lymphocytes are further differentiated to become either *B cells* (which mature in the bone marrow) or *T cells* (which travel to the thymus and mature there).

Where antibodies come from

B cells and T cells are distributed throughout the lymphoid organs, especially the lymph nodes and spleen. T cells and B cells don't attach pathogens themselves but instead produce molecules called antibodies. The antibodies attach pathogens or direct other cells, such as phagocytes, to attach them.

I'm a multi-potential cell. That means I'm capable of wearing many different hats.

Thymus

In a fetus and an infant, the *thymus* is a two-lobed mass of lymphoid tissue that is located over the base of the heart in the mediastinum. It helps form T lymphocytes for several months after birth. After this time, it has no function in the body's immunity. It gradually undergoes atrophy until only a remnant persists in adults.

Basic training

In the thymus, T cells undergo a process called T-cell education, in which the cells are "trained" to recognize other cells from the same body (self cells) and distinguish them from all other cells (nonself cells).

Peripheral lymphoid organs and tissues

Peripheral structures include the lymph nodes, lymph, the lymphatic vessels, and the spleen.

Memory jogger

To help you remember where lymphocytes mature, remember that B cells mature in the **B**one marrow and T cells mature in the **T**hymus.

Lymph nodes

The *lymph nodes* are small, oval-shaped structures located along a network of *lymph channels*. Most abundant in the head, neck, axillae, abdomen, pelvis, and groin, they help remove and destroy *antigens* (substances capable of triggering an immune response) that circulate in the blood and lymph.

Fully furnished compartments

Each lymph node is enclosed by a fibrous capsule from which bands of connective tissue extend into the node and divide it into three compartments:
• The *superficial cortex* contains follicles made up predominantly of B cells.
• The *deep cortex* and interfollicular areas consist mostly of T cells.
• The *medulla* contains numerous plasma cells that actively secrete *immunoglobulins*.

Lymph and lymphatic vessels

Lymph is a clear fluid that bathes the body tissues. It contains a liquid portion, which resembles blood plasma, as well as WBCs (mostly lymphocytes and macrophages) and antigens. Collected from body tissues, lymph seeps into *lymphatic vessels* across the vessels' thin walls. (See *Lymphatic vessels and lymph nodes.*)

Carried into the cavities...

Afferent lymphatic vessels carry lymph into the *subcapsular sinus* (or cavity) of the lymph node. From here, lymph flows through cortical sinuses and smaller radial medullary sinuses. Phagocytic cells in the deep cortex and medullary sinuses attack antigens carried in lymph. The antigens also may be trapped in the follicles of the superficial cortex.

...and coming out cleansed

Cleansed lymph leaves the node through *efferent lymphatic vessels* at the *hilum* (a depression at the exit or entrance of the node). These vessels drain into *lymph node chains* that, in turn, empty into large lymph vessels, or trunks, that drain into the subclavian vein of the vascular system.

Memory jogger

Afferent means to **bear** and efferent means to **bring out**. Therefore, it's easy to remember that the afferent lymphatic vessels **bear** lymph into the sinuses and the efferent lymphatic vessels **bring it out.**

Zoom in

Lymphatic vessels and lymph nodes

Lymphatic tissues are connected by a network of thin-walled drainage channels called *lymphatic vessels*. Resembling veins, the afferent lymphatic vessels carry lymph into lymph nodes; lymph slowly filters through the node and is collected into efferent lymphatic vessels.

You can come in but you can't go out
Lymphatic capillaries are located throughout most of the body. Wider than blood capillaries, they permit interstitial fluid to flow into them but not out.

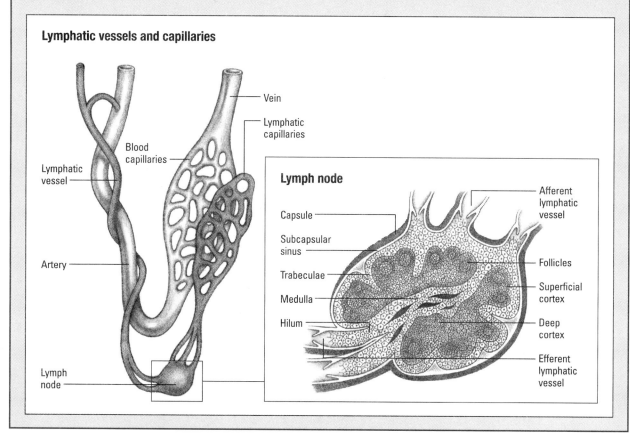

Lymphatic vessels and capillaries

- Vein
- Lymphatic capillaries
- Blood capillaries
- Lymphatic vessel
- Artery
- Lymph node

Lymph node

- Capsule
- Subcapsular sinus
- Trabeculae
- Medulla
- Hilum
- Afferent lymphatic vessel
- Follicles
- Superficial cortex
- Deep cortex
- Efferent lymphatic vessel

Getting security clearance

Usually, lymph travels through more than one lymph node because numerous nodes line the lymphatic channels that drain a particular region. For example, axillary nodes (located under the arm) filter drainage from the arms, and femoral nodes (in the inguinal region) filter drainage from the legs. This arrangement prevents organisms that enter peripheral areas from migrating unchallenged to central areas.

Spleen

Located in the left upper quadrant of the abdomen beneath the diaphragm, the *spleen* is a dark red, oval structure that is approximately the size of a fist. Bands of connective tissue from the dense fibrous capsule surrounding the spleen extend into the spleen's interior.

Pure pulp underneath

The interior, called the *splenic pulp*, contains white and red pulp. *White pulp* contains compact masses of lymphocytes surrounding branches of the splenic artery. *Red pulp* consists of a network of blood-filled *sinusoids*, supported by a framework of reticular fibers and mononuclear phagocytes, along with some lymphocytes, plasma cells, and monocytes.

A real hard worker

The spleen has several functions:
- Its phagocytes engulf and break down worn-out red blood cells (RBCs), causing the release of hemoglobin, which then breaks down into its components. These phagocytes also selectively retain and destroy damaged or abnormal RBCs and cells with large amounts of abnormal hemoglobin.
- The spleen filters and removes bacteria and other foreign substances that enter the bloodstream; these substances are promptly removed by splenic phagocytes.
- Splenic phagocytes interact with lymphocytes to initiate an immune response.
- The spleen stores blood and 20% to 30% of platelets.

> The spleen is the largest lymphatic organ because it has so many jobs to do.

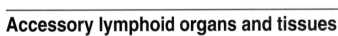

Accessory lymphoid organs and tissues

The *tonsils*, *adenoids*, *appendix*, and *Peyer's patches* remove foreign debris in much the same way lymph nodes do. They're located in food and air passages — areas where microbial access is more likely to occur.

How the immune system works

Immunity refers to the body's capacity to resist invading organisms and toxins, thereby preventing tissue and organ damage. The immune system is designed to recognize, respond to, and eliminate antigens, including bacteria, fungi, viruses, and parasites. It also preserves the body's internal environment by scavenging dead or damaged cells and patrolling for antigens.

Strategic moves

To perform these functions efficiently, the immune system uses three basic strategies:
• protective surface phenomena
• general host defenses
• specific immune responses.

Protective surface phenomena

Strategically placed physical, chemical, and mechanical barriers work to prevent the entry of potentially harmful organisms.

The forward guard

Intact and healing *skin* and *mucous membranes* provide the first line of defense against microbial invasion, preventing attachment of microorganisms. Skin *desquamation* (normal cell turnover) and low pH further impede bacterial colonization. Seromucous surfaces are protected by antibacterial substances—for instance, the enzyme *lysozyme*, found in tears, saliva, and nasal secretions.

Filtering out those that get by the first line

In the respiratory system (the easiest for microorganisms to enter), *nasal hairs* and *turbulent airflow* through the nostrils filter foreign materials. Nasal secretions contain an immunoglobulin that discourages microbe adherence. Also, a mucous layer, which is continuously sloughed off and replaced, lines the respiratory tract.

No return ticket

In the GI tract, bacteria are mechanically removed by saliva, swallowing, peristalsis, and defecation. In addition, the low pH of gastric secretions is *bactericidal* (bacteria-killing), rendering the stomach virtually free from live bacteria.

The remainder of the GI system is protected through *colonization resistance*, in which resident bacteria prevent other microorganisms from permanently making a home.

Closed neighborhood

The urinary system is sterile except for the distal end of the urethra and the urinary meatus. Urine flow, low urine pH, immunoglobulin and, in men, the bactericidal effects of *prostatic fluid* work together to impede bacterial colonization. A series of sphincters also inhibits bacterial migration.

General host defenses

When an antigen does penetrate the skin or mucous membrane, the immune system launches nonspecific cellular responses in an effort to identify and remove the invader.

Raising the red flag

The first of the nonspecific responses against an antigen, the *inflammatory response*, involves vascular and cellular changes that eliminate dead tissue, microorganisms, toxins, and inert foreign matter. (See *Understanding the inflammatory response.*)

"Phag" means *to eat* in Greek

Phagocytosis occurs after inflammation or during chronic infections. In this nonspecific response, neutrophils and macrophages engulf, digest, and dispose of the antigen.

Specific immune responses

All foreign substances elicit the same general host defenses. In addition, particular microorganisms or molecules activate specific immune responses and can involve specialized sets of immune cells. Specific responses, classified as either *humoral immunity* or *cell-mediated immunity*, are produced by lymphocytes (B cells and T cells).

I'm a resident bacterium. I live in harmony with the body without causing disease, but I keep other microorganisms from colonizing my turf.

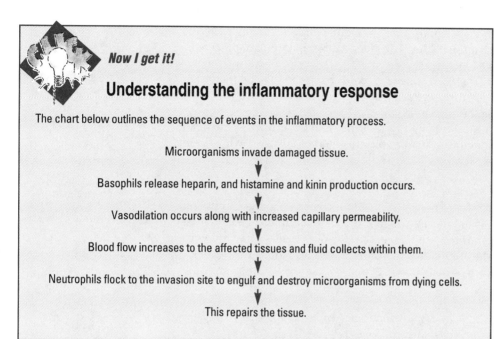

Now I get it!

Understanding the inflammatory response

The chart below outlines the sequence of events in the inflammatory process.

Microorganisms invade damaged tissue.

↓

Basophils release heparin, and histamine and kinin production occurs.

↓

Vasodilation occurs along with increased capillary permeability.

↓

Blood flow increases to the affected tissues and fluid collects within them.

↓

Neutrophils flock to the invasion site to engulf and destroy microorganisms from dying cells.

↓

This repairs the tissue.

My job is to produce antibodies. They attack the antigens for me.

Humoral immunity

In this response, an invading antigen causes B cells to divide and differentiate into plasma cells. Each plasma cell, in turn, produces and secretes large amounts of antigen-specific immunoglobulins into the bloodstream.

Five fighters

Each of the five types of *immunoglobulins* (IgA, IgD, IgE, IgG, and IgM) serves a particular function:
• IgA, IgG, and IgM guard against viral and bacterial invasion.
• IgD acts as an antigen receptor of B cells.
• IgE causes an allergic response.

I like the direct approach. I attack antigens myself.

The "Y" in how they attach to antigens

Immunoglobulins have a special molecular structure that creates a Y shape. The upper fork of the Y is designed to attach to a particular antigen; the lower stem enables the immunoglobulin to link with other structures in the immune system. Depending on the antigen, immunoglobulins can work in one of several ways:

• They can disable certain bacteria by linking with toxins that the bacteria produce; these immunoglobulins are called *antitoxins*.
• They can *opsonize* (coat) bacteria, making them targets for scavenging by phagocytosis. (See *How macrophages accomplish phagocytosis*.)
• Most commonly, they can link to antigens, causing the immune system to produce and circulate enzymes called *complement*.

A break in the action

After the body's initial exposure to an antigen, a time lag occurs during which little or no antibody can be detected. During this time, the B cell recognizes the antigen and the sequence of division, differentiation, and antibody formation begins.

You thought I was slow in the first round? Well, just wait. I always come back faster and meaner in the second.

The first response

The *primary antibody response* occurs 4 to 10 days after first-time antigen exposure. During this response, immunoglobulin levels increase, then quickly dissipate, and IgM antibodies form.

Second response: Hit 'em hard, hit 'em fast

Subsequent exposure to the same antigen initiates a *secondary antibody response*. In this response, memory B cells manufacture antibodies (now mainly IgG), achieving peak levels in 1 to 2 days. These elevated levels persist for months, then fall slowly. Thus, the secondary antibody response is faster, more intense, and more persistent than the primary response. This response intensifies with each subsequent exposure to the same antigen.

Getting complex

After the antibody reacts to the antigen, an *antigen-antibody complex* forms. The complex serves several functions. First, a macrophage processes the antigen and presents it to antigen-specific B cells. Then the antibody activates the complement system, causing an *enzymatic cascade* that destroys the antigen.

Zoom in

How macrophages accomplish phagocytosis

Microorganisms and other antigens that invade the skin and mucous membranes are removed by *phagocytosis,* a defense mechanism carried out by macrophages (mononuclear leukocytes) and neutrophils (polymorphonuclear leukocytes). Here is how macrophages accomplish phagocytosis.

Chemotaxis
Chemotactic factors attract macrophages to the antigen site.

Microorganism

Chemotactic factors

Macrophage

Opsonization
Antibody (immunoglobulin G) or complement fragment coats the microorganism, enhancing macrophage binding to the antigen, now called an *opsinogen.*

Opsonized microorganism

Ingestion
The macrophage extends its membrane around the opsonized microorganism, engulfing it within a vacuole *(phagosome).*

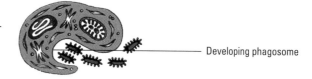

Developing phagosome

Digestion
As the phagosome shifts away from the cell periphery, it merges with lysosomes, forming a *phagolysosome,* where antigen destruction occurs.

Phagolysosome

Release
When digestion is complete, the macrophage expels digestive debris, including lysosomes, prostaglandins, complement components, and interferon, which continue to mediate the immune response.

Digestive debris

Complement system

The activated *complement system*, which bridges humoral and cell-mediated immunity, attracts phagocytic neutrophils and macrophages to the antigen site.

The complement cascade plays a crucial role in the inflammatory response.

Working together

Indispensable to the humoral immune response, the complement system consists of about 25 enzymes that "complement" the work of antibodies by aiding phagocytosis or destroying bacteria cells (through puncture of their cell membranes).

A ripple effect

Complement proteins travel in the bloodstream in an inactive form. When the first complement substance is triggered (typically by an antibody interlocked with an antigen), it sets in motion a ripple effect. As each component is activated in turn, it acts on the next component in a sequence of carefully controlled steps called the *complement cascade*.

Attack mode

This cascade leads to the creation of the *membrane attack complex*. Inserted into the membrane of the target cell, this complex creates a channel through which fluids and molecules flow in and out. The target cell then swells and eventually bursts.

Other benefits flow from the complement cascade

By-products of the complement cascade also produce:
• the inflammatory response (resulting from release of the contents of mast cells and basophils)
• stimulation and attraction of neutrophils (which participate in phagocytosis)
• coating of target cells by *C3b* (an inactivated fragment of the complement protein C3), making them attractive to phagocytes.

Cell-mediated immunity

Cell-mediated immunity protects the body against bacterial, viral, and fungal infections and provides resistance against transplanted cells and tumor cells.

Ever vigilant

In this immune response, a macrophage processes the antigen, which is then presented to T cells. Some T cells become sensitized and destroy the antigen; others release *lymphokines*, which activate macrophages that destroy the antigen. Sensitized T cells then travel through the blood and lymphatic systems, providing ongoing surveillance in their quest for specific antigens.

When the immune system fails

Because of their complexity, the processes involved in host defense and immune response may malfunction. When the body's defenses are exaggerated, misdirected, or either absent or depressed, the result may be a *hypersensitivity disorder, autoimmunity,* or *immunodeficiency,* respectively.

> Hypersensitivity stems from an exaggerated or inappropriate immune response.

Hypersensitivity disorders

An exaggerated or inappropriate immune response may lead to various hypersensitivity disorders.

Typing them out

Such disorders are classified as type I through type IV, depending on which immune system activity causes tissue damage, although some overlap exists:
- Type I disorders are *anaphylactic (immediate, atopic, IgE-mediated reaginic) reactions.* Examples of type I disorders include systemic anaphylaxis, hay fever (seasonal allergic rhinitis), reactions to insect stings, some food and drug reactions, some cases of urticaria, and infantile eczema.
- Type II disorders are *cytotoxic (cytolytic, complement-dependent cytotoxicity) reactions.* Examples of type II disorders include Goodpasture's syndrome, autoimmune hemolytic anemia, transfusion reactions, hemolytic disease of the newborn, myasthenia gravis, and some drug reactions.
- Type III disorders are *immune complex disease reactions.* Examples of type III disorders are reactions associated with such infections as hepatitis B and bacterial endocarditis; cancers, in which a serum sickness-like

syndrome may occur; and autoimmune disorders such as systemic lupus erythematosus. This hypersensitivity reaction may also follow drug or serum therapy.

• Type IV disorders are *delayed (cell-mediated) hypersensitivity reactions*. Type IV disorders include tuberculin reactions, contact hypersensitivity, and sarcoidosis.

Autoimmune disorders

Autoimmune disorders are marked by an abnormal response to one's own tissue.

Diffusion can lead to confusion

Autoimmunity leads to a sequence of tissue reactions and damage that may produce diffuse systemic signs and symptoms. Among the autoimmune disorders are rheumatoid arthritis, juvenile rheumatoid arthritis, psoriatic arthritis, ankylosing spondylitis, Sjögren's syndrome, and lupus erythematosus.

Immunodeficiency

Immunodeficiency disorders are caused by an absent or depressed immune response in various forms.

Unfortunately, no deficiency of immunodeficiency disorders

Immunodeficiency disorders include X-linked infantile hypogammaglobulinemia, common variable immunodeficiency, DiGeorge's syndrome, acquired immunodeficiency syndrome, chronic granulomatous disease, ataxia-telangiectasia, severe combined immunodeficiency disease, and complement deficiencies.

Immunodeficiency disorders occur when the immune system is depressed or on a downward slide.

Quick quiz

1. Stem cells are multipotential and develop into other types of cells, such as blood cells and immune system cells, through the process of:

    A.   differentiation.
    B.   phagocytosis.
    C.   hematopoiesis.

Answer: C. Hematopoiesis is the formation of blood cells, which occurs in the bone marrow.

2. Lymph nodes help remove foreign debris and antigens from the body by phagocytosis. Cleansed lymph leaves the lymph node through:

    A.   afferent lymphatic vessels.
    B.   efferent lymphatic vessels.
    C.   lymphatic capillaries.

Answer: B. Efferent lymphatic vessels drain into lymph node chains, then into large lymph vessels and, finally, into the subclavian vein.

3. An invading antigen would stay away from all of the following except:

    A.   immunoglobulins.
    B.   neutrophils.
    C.   normal red blood cells.

Answer: C. Immunoglobulins and neutrophils are involved in the process of phagocytosis.

4. If you were a microbe trying to invade the body, which unprotected body part would be the easiest to enter?

    A.   Skin
    B.   Respiratory tract
    C.   Urinary tract

Answer: B. The respiratory tract is the most easily entered and needs protection.

5. If your child were allergic to peanut butter and he ate five peanut butter cookies, which of the following antigen-specific immunoglobulins would be produced in his body?

     A.   IgA
     B.   IgD
     C.   IgE

Answer: C. IgE is responsible for allergic reactions.

Scoring

☆☆☆   If you answered all five questions correctly, impressive! You've responded like a pro to all of our immunologic challenges!

☆☆   If you answered three or four questions correctly, prodigious! You've proved your multi-potential!

☆   If you answered fewer than three questions correctly, you have an immunodeficiency. Take some vitamin C, and read this chapter again in the morning!

Respiratory system

Just the facts

In this chapter, you'll learn:

♦ structures of the respiratory system and their function

♦ how inspiration and expiration occur

♦ how gas exchange takes place in the alveoli

♦ how problems with the nervous, musculoskeletal, and pulmonary systems can affect breathing

♦ role of the lungs in acid-base balance.

The respiratory system at a glance

The respiratory system consists of the:

 upper respiratory tract

 lower respiratory tract

 lungs

 the thoracic cavity.

This system maintains the exchange of oxygen and carbon dioxide in the lungs and tissues. The respiratory system also helps regulate the body's acid-base balance.

Upper respiratory tract

The upper respiratory tract consists primarily of the nose, mouth, nasopharynx, oropharynx, laryngopharynx, and larynx. These structures warm and humidify inspired

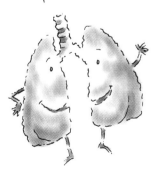

The respiratory system maintains the exchange of oxygen and carbon dioxide in the lungs and tissues.

air. They're also responsible for taste, smell, and the chewing and swallowing of food. (See *Structures of the respiratory system.*)

Nostrils and nasal passages

Air enters the body through the nostrils (*nares*). In the nares, small hairs known as *vibrissae* filter out dust and large foreign particles. Air then passes into the two nasal passages, which are separated by the *septum*. Cartilage forms the anterior walls of the nasal passages; bony structures (*conchae* or *turbinates*) form the posterior walls.

Just passing through

The *conchae* warm and humidify air before it passes into the nasopharynx. Their mucus layer also traps finer foreign particles, which the cilia carry to the pharynx to be swallowed.

Sinuses and pharynx

The four paranasal sinuses are located in the frontal, sphenoid, and maxillary bones. The sinuses provide speech resonance.

Air passes from the nasal cavity into the muscular nasopharynx through the *choanae*, a pair of posterior openings in the nasal cavity that remain constantly open.

Oropharynx and laryngopharynx

The oropharynx is the posterior wall of the mouth. It connects the nasopharynx and the laryngopharynx. The laryngopharynx extends to the esophagus and larynx.

Larynx

The larynx contains the vocal cords and connects the pharynx with the trachea. Muscles and cartilage form the walls of the larynx, including the large, shield-shaped thyroid cartilage situated just under the jaw line.

These involuntary defense mechanisms help protect the respiratory system from infection and prevent foreign-body inhalation.

- Sneezing
- Coughing
- Gagging
- Spasms

Body shop

Structures of the respiratory system

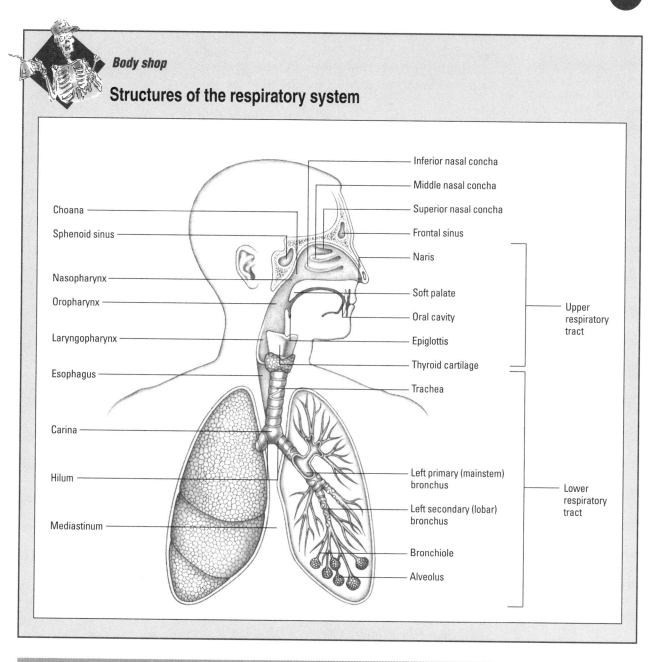

Choana

Sphenoid sinus

Nasopharynx

Oropharynx

Laryngopharynx

Esophagus

Carina

Hilum

Mediastinum

Inferior nasal concha

Middle nasal concha

Superior nasal concha

Frontal sinus

Naris

Soft palate

Oral cavity

Epiglottis

Thyroid cartilage

Trachea

Left primary (mainstem) bronchus

Left secondary (lobar) bronchus

Bronchiole

Alveolus

Upper respiratory tract

Lower respiratory tract

Lower respiratory tract

The lower respiratory tract consists of the trachea, bronchi, and lungs. Functionally, the lower tract is subdivided into the *conducting airways* and the *acinus*. The acinus serves as the area of gas exchange. A mucous membrane contains hairlike cilia and lines the lower tract. Cilia constantly clean the tract and carry foreign matter upward for swallowing or expectoration.

Conducting airways

The conducting airways, which contain the trachea and bronchi, help facilitate gas exchange.

Trachea

The trachea extends from the *cricoid cartilage* at the top to the carina (also called the *tracheal bifurcation*). The carina is a ridge-shaped structure at the level of the sixth or seventh thoracic vertebra. C-shaped cartilage rings reinforce and protect the trachea to prevent it from collapsing.

Bronchi

The primary bronchi begin at the carina. The right mainstem bronchus — shorter, wider, and more vertical than the left — supplies air to the right lung. The left mainstem bronchus delivers air to the left lung.

Secondary bronchi and the hilum

The mainstem bronchi divide into the five lobar bronchi (secondary bronchi). Along with blood vessels, nerves, and lymphatics, the secondary bronchi enter the pleural cavities and the lungs at the *hilum*. Located behind the heart, the hilum is a slit on the lung's medial surface where the lungs are anchored.

Bronchi branches out

Each lobar bronchus enters a lobe in each lung. Within its lobe, each of the lobar bronchi branches into segmental bronchi (tertiary bronchi). The segments continue to branch into smaller and smaller bronchi, finally branching into bronchioles.

The larger bronchi consist of cartilage, smooth muscle, and epithelium. As the bronchi become smaller, they lose first cartilage, then smooth muscle until, finally, the smallest bronchioles consist of just a single layer of epithelial cells.

The function of the lower respiratory tract is divided between the conducting airways and the acinus.

Acinus

Each bronchiole includes terminal bronchioles and the acinus — the chief respiratory unit for gas exchange. (See *A close look at a pulmonary airway.*)

Respiratory bronchioles

Within the acinus, terminal bronchioles branch into yet smaller respiratory bronchioles. The respiratory bronchioles feed directly into alveoli at sites along their walls.

Alveolar walls contain two basic epithelial cell types:

👉 Type I cells are the most abundant. These cells are thin, flat, squamous cells across which gas exchange occurs.

The smallest of the bronchioles are composed of just a single layer of cells.

Zoom in

A close look at a pulmonary airway

As illustrated below, each lobule contains terminal bronchioles and the acinus, consisting of respiratory bronchioles and alveolar sacs.

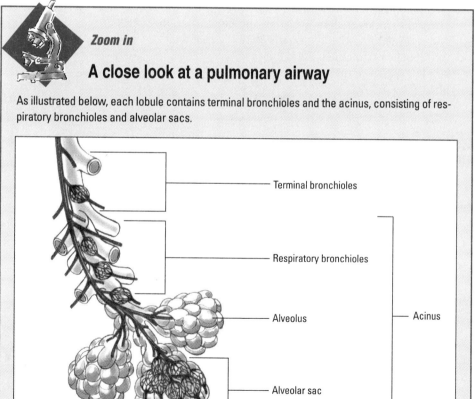

Terminal bronchioles

Respiratory bronchioles

Alveolus

Acinus

Alveolar sac

Type II cells secrete *surfactant*, a substance that coats the alveolus and promotes gas exchange by lowering surface tension.

Alveoli

The respiratory bronchioles eventually become alveolar ducts, which terminate in clusters of capillary-swathed alveoli called alveolar sacs. Gas exchange takes place through the alveoli.

Lungs and accessory structures

The cone-shaped lungs hang suspended in the right and left pleural cavities, straddling the heart, and anchored by root and pulmonary ligaments.

The right lung is shorter, broader, and larger than the left. It has three lobes and handles 55% of gas exchange. The left lung has two lobes. Each lung's concave base rests on the diaphragm; the apex extends about ½″ (1.3 cm) above the first rib.

Pleura and pleural cavities

The pleura — the membrane that totally encloses the lung — is composed of a visceral layer and a parietal layer. The visceral pleura hugs the entire lung surface, including the areas between the lobes. The parietal pleura lines the inner surface of the chest wall and upper surface of the diaphragm.

Serous fluid has serious functions

The pleural cavity — the tiny area between the visceral and parietal pleural layers — contains a thin film of serous fluid. This fluid has two functions:

It lubricates the pleural surfaces so that they slide smoothly against each other as the lungs expand and contract.

It creates a bond between the layers that causes the lungs to move with the chest wall during breathing.

I hang in the right and left pleural cavities.

Thoracic cavity

The thoracic cavity is an area that is surrounded by the diaphragm (below), the scalene muscles and fasciae of the neck (above), and the ribs, intercostal muscles, vertebrae, sternum, and ligaments (around the circumference.)

Mediastinum

The space between the lungs is called the *mediastinum*. It contains:
- the heart and pericardium
- the thoracic aorta
- the pulmonary artery and veins
- the venae cavae and azygos veins
- the thymus, lymph nodes, and vessels
- the trachea, esophagus, and thoracic duct
- the vagus, cardiac, and phrenic nerves.

Thoracic cage

The thoracic cage is composed of bone and cartilage. It supports and protects the lungs, allowing them to expand and contract.

Posterior thoracic cage

The vertebral column and 12 pairs of ribs form the posterior portion of the thoracic cage. The ribs form the major portion of the thoracic cage. They extend from the thoracic vertebrae toward the anterior thorax.

Anterior thoracic cage

The anterior thoracic cage consists of the manubrium, sternum, xiphoid process, and ribs. It protects the mediastinal organs that lie between the right and left pleural cavities.

Rib numbers

Ribs 1 through 7 attach directly to the sternum; ribs 8 through 10 attach to the cartilage of the preceding rib. The other 2 pairs of ribs are "free-floating" — they don't

The ribs, like the vertebrae, are numbered from top to bottom.

Body shop

Locating lung structures in the thoracic cage

The ribs, vertebrae, and other structures of the thoracic cage act as landmarks that you can use to identify underlying structures.

From an anterior view

- The base of each lung rests at the level of the sixth rib at the midclavicular line and the eighth rib at the midaxillary line.
- The apex of each lung extends about ¾" to 1½" (1.9 to 3.8 cm) above the inner aspects of the clavicles.
- The upper lobe of the right lung ends level with the fourth rib at the midclavicular line and with the fifth rib at the midaxillary line.
- The middle lobe of the right lung extends triangularly from the fourth to the sixth rib at the midclavicular line and to the fifth rib at the midaxillary line.
- Because the left lung doesn't have a middle lobe, the upper lobe of the left lung ends level with the fourth rib at the midclavicular line and with the fifth rib at the midaxillary line.

From a posterior view

- The lungs extend from the cervical area to the level of the tenth thoracic vertebra (T10). On deep inspiration, the lungs may descend to T12.
- An imaginary line, stretching from the T3 level along the inferior border of the

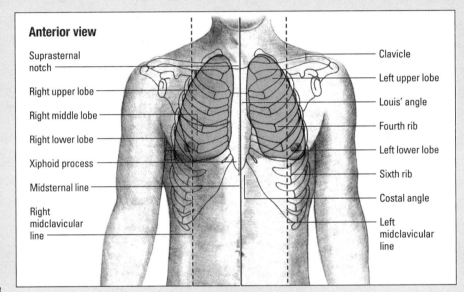

Anterior view

Suprasternal notch
Right upper lobe
Right middle lobe
Right lower lobe
Xiphoid process
Midsternal line
Right midclavicular line

Clavicle
Left upper lobe
Louis' angle
Fourth rib
Left lower lobe
Sixth rib
Costal angle
Left midclavicular line

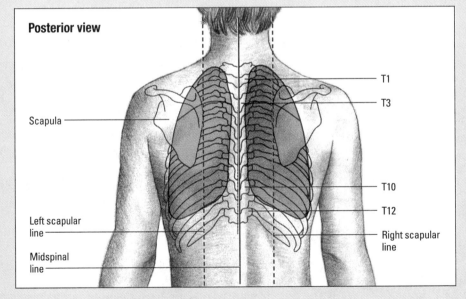

Posterior view

Scapula
Left scapular line
Midspinal line

T1
T3
T10
T12
Right scapular line

Locating lung structures in the thoracic cage (continued)

scapulae to the fifth rib at the midaxillary line, separates the upper lobes of both lungs.
- The upper lobes exist above T3; the lower lobes exist below T3 and extend to the level of T10.
- The diaphragm originates around the ninth or tenth rib.

From a lateral view

The right and left lateral rib cages cover the lobes of the right and left lungs, respectively. Beneath these structures, the lungs extend from just above the clavicles to the level of the eighth rib. The left lateral thorax allows access to two lobes; the right lateral thorax, to three lobes.

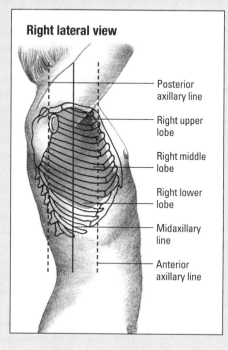

Right lateral view

Posterior axillary line

Right upper lobe

Right middle lobe

Right lower lobe

Midaxillary line

Anterior axillary line

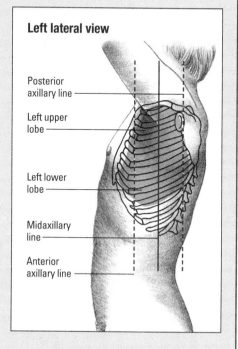

Left lateral view

Posterior axillary line

Left upper lobe

Left lower lobe

Midaxillary line

Anterior axillary line

attach to any part of the anterior thoracic cage. Rib 11 ends anterolaterally, and rib 12 ends laterally.

Bordering on the costal angle

The lower parts of the rib cage (costal margins) near the xiphoid process form the borders of the costal angle — an angle of about 90 degrees in a normal person. (See *Locating lung structures in the thoracic cage.*)

It's suprasternal

Above the anterior thorax is a depression called the suprasternal notch. Because the suprasternal notch isn't covered by the rib cage like the rest of the thorax, the trachea and aortic pulsation can be palpated here.

Inspiration and expiration

Breathing involves two actions: inspiration (an active process) and expiration (a relatively passive process). Both actions rely on respiratory muscle function and the effects of pressure differences in the lungs.

It's perfectly normal!

During normal respiration, the external intercostal muscles aid the diaphragm, the major muscle of respiration. The diaphragm descends to lengthen the chest cavity, while the external intercostal muscles (located between and along the lower borders of the ribs) contract to expand the anteroposterior diameter. This coordinated action causes inspiration. Rising of the diaphragm and relaxation of the intercostal muscles causes expiration. (See *Mechanics of respiration.*)

Forced inspiration and active expiration

During exercise, when the body needs increased oxygenation, or in certain disease states that require forced inspiration and active expiration, the accessory muscles of respiration also participate.

Forced inspiration

During forced inspiration:
• the pectoral muscles (upper chest) raise the chest to increase the anteroposterior diameter
• the sternocleidomastoid muscles (side of neck) raise the sternum
• the scalene muscles (in the neck) elevate, fix, and expand the upper chest
• the posterior trapezius muscles (upper back) raise the thoracic cage.

Active expiration

During active expiration, the internal intercostal muscles contract to shorten the chest's transverse diameter and the abdominal rectus muscles pull down the lower chest, thus depressing the lower ribs.

Body shop

Mechanics of respiration

The muscles of respiration help the chest cavity expand and contract. Pressure differences between atmospheric air and the lungs help produce air movement. These illustrations show the muscles that work together to allow inspiration and expiration.

Anterior view

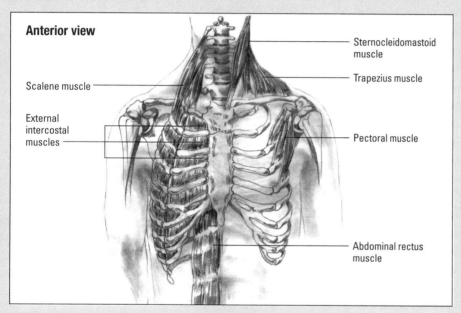

Scalene muscle

External intercostal muscles

Sternocleidomastoid muscle

Trapezius muscle

Pectoral muscle

Abdominal rectus muscle

Posterior view

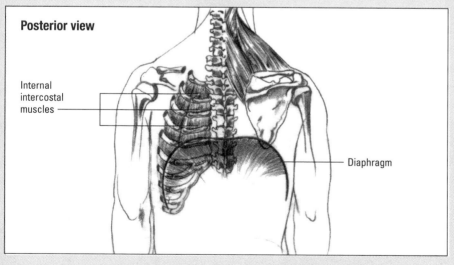

Internal intercostal muscles

Diaphragm

Now I get it!

Tracing pulmonary circulation

The right and left pulmonary arteries carry deoxygenated blood from the right side of the heart to the lungs. These arteries divide into distal branches, called arterioles, which eventually terminate as a concentrated capillary network in the alveoli and alveolar sac, where gas exchange occurs.

Venules — the end branches of the pulmonary veins — collect oxygenated blood from the capillaries and transport it to larger vessels, which in turn lead to the pulmonary veins. The pulmonary veins enter the left side of the heart and distribute oxygenated blood throughout the body.

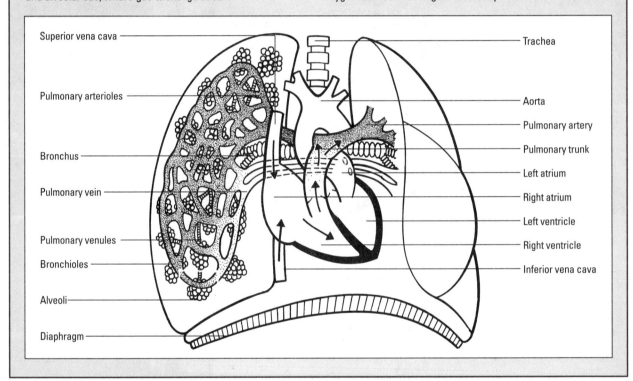

Gas station

Oxygen-depleted blood enters the lungs from the pulmonary artery of the heart's right ventricle, then flows through the main pulmonary arteries into the smaller vessels of the pleural cavities and the main bronchi, through the arterioles and, eventually, to the capillary networks in the alveoli. (See *Tracing pulmonary circulation.*) In the alveoli, gas exchange — oxygen and carbon dioxide diffusion — takes place.

External and internal respiration

Effective respiration consists of gas exchange in the lungs, called *external respiration,* and gas exchange in the tissues, called *internal respiration.*

External respiration occurs through three processes:

ventilation — gas distribution into and out of the pulmonary airways

pulmonary perfusion — blood flow from the right side of the heart, through the pulmonary circulation, and into the left side of the heart

diffusion — gas movement through a semipermeable membrane from an area of greater concentration to one of lesser concentration.

Internal respiration occurs only through diffusion.

Ventilation

Ventilation is the distribution of gases (oxygen and carbon dioxide) into and out of the pulmonary airways. Problems within the nervous, musculoskeletal, and pulmonary systems cause breathing effectiveness to be greatly compromised.

Nervous system influence

Involuntary breathing results from stimulation of the respiratory center in the medulla and the pons of the brain. The medulla controls the rate and depth of respiration; the pons moderates the rhythm of the switch from inspiration to expiration.

Musculoskeletal influence

The adult thorax is flexible — its shape can be changed by contracting the chest muscles. The medulla controls ventilation primarily by stimulating contraction of the diaphragm and external intercostal muscles. These actions produce the intrapulmonary pressure changes that cause inspiration.

Pulmonary influence

Airflow distribution can be affected by many factors:
• airflow pattern (see *Comparing airflow patterns,* page 174)

> The central nervous system's respiratory center is located in the lateral medulla.

- volume and location of the functional reserve capacity (air retained in the alveoli that prevents their collapse during respiration)
- degree of intrapulmonary resistance
- presence of lung disease.

The path of least resistance

If airflow is disrupted for any reason, airflow distribution follows the path of least resistance.

Increased workload, decreased efficiency

Other musculoskeletal and intrapulmonary factors can affect airflow and, in turn, may affect breathing. For instance, forced breathing, as in emphysema, activates accessory muscles of respiration, which require additional

Comparing airflow patterns

The pattern of airflow through the respiratory passages affects airway resistance.

Laminar flow
Laminar flow, a linear pattern that occurs at low flow rates, offers minimal resistance. This flow type occurs mainly in the small peripheral airways of the bronchial tree.

Turbulent flow
The eddying pattern of turbulent flow creates friction and increases resistance. Turbulent flow is normal in the trachea and large central bronchi. If the smaller airways become constricted or clogged with secretions, however, turbulent flow may also occur there.

Transitional flow
A mixed pattern known as transitional flow is common at lower flow rates in the larger airways, especially where the airways narrow from obstruction, meet, or branch.

Laminar flow	Turbulent flow	Transitional flow

oxygen to work. This results in less efficient ventilation with an increased workload.

Airflow interference and alterations

Other airflow alterations can also increase oxygen and energy demand and cause respiratory muscle fatigue. These conditions include interference with expansion of the lungs or thorax (changes in compliance) and interference with airflow in the tracheobronchial tree (changes in resistance).

Pulmonary perfusion

Pulmonary perfusion refers to blood flow from the right side of the heart, through the pulmonary circulation, and into the left side of the heart. Perfusion aids external respiration. Normal pulmonary blood flow allows alveolar gas exchange, but many factors may interfere with gas transport to the alveoli. Here are some examples:
• Cardiac output less than the average of 5 L/minute decreases gas exchange by reducing blood flow.
• Elevations in pulmonary and systemic resistance reduce blood flow.
• Abnormal or insufficient hemoglobin picks up less oxygen for exchange.

Ventilation-perfusion match

Gravity can affect oxygen and carbon dioxide transport in a positive way. Gravity causes more unoxygenated blood to travel to the lower and middle lung lobes than to the upper lobes.

This explains why ventilation and perfusion differ in the various parts of the lungs. Areas where perfusion and ventilation are similar have what is referred to as a *ventilation-perfusion match;* in such areas, gas exchange is most efficient. (See *What happens in ventilation-perfusion mismatch,* page 176.)

Optimal conditions for diffusion

In diffusion, oxygen and carbon dioxide molecules move between the alveoli and capillaries. The direction of movement is always from an area of greater concentration to one of lesser concentration. In the process, oxygen moves across the alveolar and capillary membranes, dissolves in the plasma, and then passes through the red

What happens in ventilation-perfusion mismatch

Ideally, the amount of air in the alveoli (a reflection of ventilation) matches the amount of blood in the capillaries (a reflection of perfusion). This allows gas exchange to proceed smoothly.

   This ventilation-perfusion (V/Q) ratio is actually unequal: The alveoli receive air at a rate of approximately 4 L/minute, while the capillaries supply blood at a rate of about 5 L/minute. This creates a V/Q mismatch of 4:5 or 0.8.

Normal

In the normal lung, ventilation closely matches perfusion.

Shunt

Perfusion without ventilation usually results from airway obstruction, particularly that caused by acute diseases, such as atelectasis and pneumonia.

Dead-space ventilation

Normal ventilation without perfusion usually results from a perfusion defect such as pulmonary embolism.

Silent unit

Absence of ventilation and perfusion usually stems from multiple causes such as pulmonary embolism with resultant adult respiratory distress syndrome and emphysema.

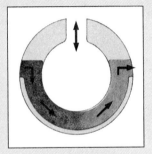 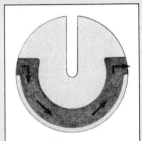 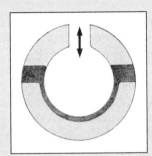 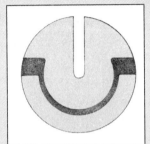

blood cell (RBC) membrane. Carbon dioxide moves in the opposite direction.

Spaces in-between

The epithelial membranes lining the alveoli and capillaries must be intact. Both the alveolar epithelium and the capillary endothelium are composed of a single layer of cells. Between these layers are tiny interstitial spaces filled with elastin and collagen.

From the RBCs to the alveoli

Normally, oxygen and carbon dioxide move easily through all of these layers. Oxygen moves from the alveoli into the bloodstream, where it's taken up by hemoglobin in the RBCs. When it arrives in the bloodstream, it displaces carbon dioxide (the by-product of metabolism), which diffuses from RBCs into the blood and then to the alveoli.

To bind or not to bind

Most transported oxygen binds with hemoglobin to form oxyhemoglobin, but a small portion dissolves in the plasma. The portion of oxygen that dissolves in plasma can be measured as the partial pressure of oxygen in arterial blood, or Pao_2.

After oxygen binds to hemoglobin, RBCs travel to the tissues. Through cellular diffusion, internal respiration occurs when RBCs release oxygen and absorb carbon dioxide. The RBCs then transport the carbon dioxide back to the lungs for removal during expiration. (See *Exchanging gases*, page 178.)

During diffusion, oxygen and carbon dioxide travel the same path but in opposite directions — from an area of greater concentration to one of lesser concentration.

Acid-base balance

Oxygen taken up in the lungs is transported to the tissues by the circulatory system, which exchanges it for carbon dioxide produced by metabolism in body cells. Because carbon dioxide is more soluble than oxygen, it dissolves in the blood. In the blood, most of the oxygen forms bicarbonate (base) and smaller amounts form carbonic acid (acid).

Respiratory responses

The lungs control bicarbonate levels by converting bicarbonate to carbon dioxide and water for excretion. In response to signals from the medulla, the lungs can change the rate and depth of breathing. This change allows for adjustments in the amount of carbon dioxide lost to help maintain acid-base balance.

Metabolic alkalosis

For example, in *metabolic alkalosis* (a condition resulting from excess bicarbonate retention), the rate and depth of ventilation decrease so that carbon dioxide can be retained; this increases carbonic acid levels.

Metabolic acidosis

In *metabolic acidosis* (a condition resulting from excess acid retention or excess bicarbonate loss), the lungs in-

Now I get it!

Exchanging gases

Gas exchange occurs very rapidly in the millions of tiny, thin-membraned alveoli within the respiratory units. Inside these air sacs, oxygen from inhaled air diffuses into the blood while carbon dioxide diffuses from the blood into the air and is exhaled. Blood then circulates throughout the body, delivering oxygen and picking up carbon dioxide. Finally, the blood returns to the lungs to be oxygenated again.

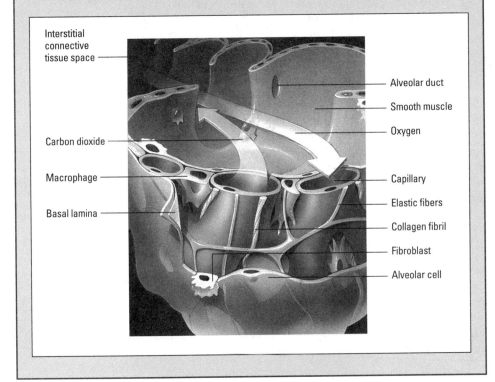

- Interstitial connective tissue space
- Carbon dioxide
- Macrophage
- Basal lamina
- Alveolar duct
- Smooth muscle
- Oxygen
- Capillary
- Elastic fibers
- Collagen fibril
- Fibroblast
- Alveolar cell

It takes cooperation! I send the messages...

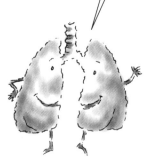

...and I change the rate and depth of breathing. Together we adjust levels of carbon dioxide and maintain acid-base balance.

crease the rate and depth of ventilation to eliminate excess carbon dioxide, thus reducing carbonic acid levels.

Broken balance beam

When the lungs don't function properly, an acid-base imbalance results. For example, they can cause respiratory acidosis through *hypoventilation* (reduced rate and depth of ventilation), which leads to carbon dioxide retention.

Quick quiz

1. Which of the following is the chief respiratory unit for gas exchange?
  A. Acinus
  B. Alveoli
  C. Terminal bronchioles

Answer: A. The acinus is the chief respiratory unit for gas exchange.

2. How many lobes does the right lung have?
  A. Six
  B. Two
  C. Three

Answer: C. The right lung has three lobes.

3. During gas exchange, oxygen and carbon dioxide diffusion occurs in the:
  A. venules.
  B. alveoli.
  C. red blood cells.

Answer: B. Oxygen and carbon dioxide diffusion occurs in the alveoli.

4. When oxygen passes through the alveoli into the bloodstream, it binds with hemoglobin to form which of the following?
  A. Red blood cells
  B. Carbon dioxide
  C. Oxyhemoglobin

Answer: C. When oxygen passes through the alveoli into the bloodstream, it binds with hemoglobin to form oxyhemoglobin.

Scoring

☆☆☆   If you answered all four questions correctly, extraordinary! Take a deep breath! You've responded extremely well to the respiratory system.

☆☆   If you answered two or three questions correctly, fascinating! You're breezing through these systems like a whirlwind.

☆   If you answered fewer than two questions correctly, get inspired! There are six more quick quizzes to go!

Gastrointestinal system

Just the facts

In this chapter, you'll learn:

♦ two major components of the GI system

♦ phases of digestion

♦ function of GI hormones

♦ sites and mechanisms of gastric secretions.

The GI system at a glance

The GI system has two major components: the *alimentary canal* (also called the GI tract) and the *accessory GI organs*.

The GI tract serves two major functions:

• *digestion,* or the breaking down of food and fluid into simple chemicals that can be absorbed into the bloodstream and transported throughout the body

• *elimination* of waste products through excretion of stool.

Digestion and excretion of waste products are the GI tract's major functions.

Alimentary canal

The alimentary canal is a hollow muscular tube that begins in the mouth and extends to the anus. It includes the pharynx, esophagus, stomach, small intestine, and large intestine. (See *Structures of the GI system*, page 182.)

Body shop

Structures of the GI system

The GI system includes the alimentary canal (pharynx, esophagus, stomach, and small and large intestines) and the accessory organs (liver, biliary duct system, and pancreas).

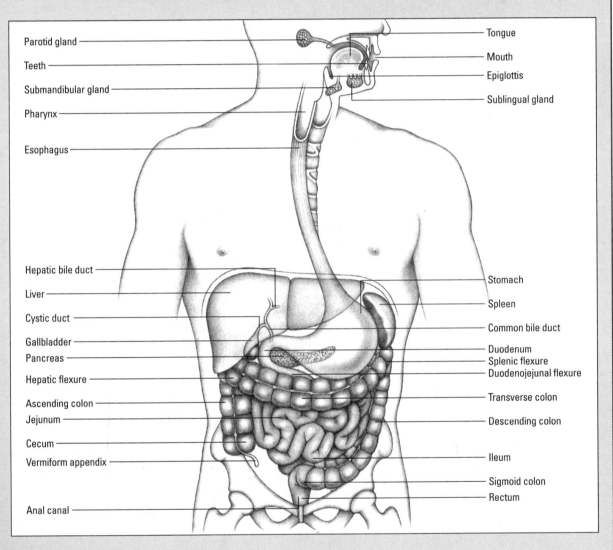

Parotid gland
Teeth
Submandibular gland
Pharynx
Esophagus
Hepatic bile duct
Liver
Cystic duct
Gallbladder
Pancreas
Hepatic flexure
Ascending colon
Jejunum
Cecum
Vermiform appendix
Anal canal

Tongue
Mouth
Epiglottis
Sublingual gland
Stomach
Spleen
Common bile duct
Duodenum
Splenic flexure
Duodenojejunal flexure
Transverse colon
Descending colon
Ileum
Sigmoid colon
Rectum

Mouth

The mouth (also called the *buccal cavity* or *oral cavity*) is bounded by the lips (labia), cheeks, palate (roof of the mouth), and tongue and contains the teeth. Ducts connect the mouth with the three major pairs of salivary glands:

 parotid

 submandibular

 sublingual.

These glands secrete saliva to moisten food during chewing. The mouth initiates the mechanical breakdown of food. (See *Oral cavity*, page 184.)

Pharynx

The *pharynx* is a cavity that extends from the base of the skull to the esophagus. The pharynx aids swallowing by grasping food and propelling it toward the esophagus.

Esophagus

The *esophagus* is a muscular tube that extends from the pharynx through the mediastinum to the stomach.

Peristalsis propels liquids and solids through the esophagus into the stomach. The cricopharyngeal sphincter — a sphincter at the upper border of the esophagus — must relax for food to enter the esophagus.

Swallowing triggers the passage of food from the pharynx to the esophagus.

Stomach

The *stomach* is a collapsible, pouchlike structure in the left upper part of the abdominal cavity, just below the diaphragm. It lies in the left side of the superior part of the abdominal cavity, its upper border attaching to the lower end of the esophagus. The lateral surface of the stomach is called the *greater curvature;* the medial surface, the *lesser curvature.*

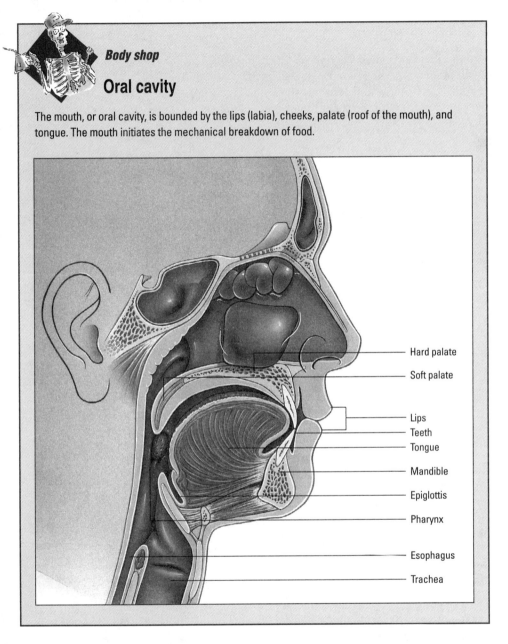

Body shop

Oral cavity

The mouth, or oral cavity, is bounded by the lips (labia), cheeks, palate (roof of the mouth), and tongue. The mouth initiates the mechanical breakdown of food.

Hard palate

Soft palate

Lips

Teeth

Tongue

Mandible

Epiglottis

Pharynx

Esophagus

Trachea

Size does matter

The size of the stomach varies with the degree of distention. Overeating can cause marked distention, pushing on the diaphragm and causing shortness of breath. (See *Stomach, duodenum, and jejunum.*)

Zoom in

Stomach, duodenum, and jejunum

This cross section of the stomach shows the G-cells (which secrete gastrin) in the pyloric glands. The cross section of the duodenum and jejunum show the S-cells (which secrete secretin) in the duodenal and jejunal glands.

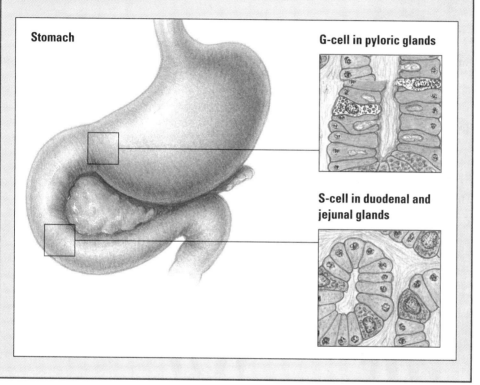

Stomach

G-cell in pyloric glands

S-cell in duodenal and jejunal glands

Four regions

The stomach has four main regions:

The *cardia* lies near the junction of the stomach and esophagus.

The *fundus* is an enlarged portion above and to the left of the esophageal opening into the stomach.

The *body* is the middle portion of the stomach.

The *pylorus* is the lower portion, lying near the junction of the stomach and duodenum.

What am I doing here?

The stomach has several functions. It:
- serves as a temporary storage area for food
- begins digestion
- breaks down food into *chyme*, a semifluid substance
- moves the gastric contents into the small intestine.

Small intestine

The *small intestine* is a tube that measures about 20′ (6 m) long. It's the longest organ of the GI tract and has three major divisions:
- The *duodenum* is the longest and most superior division.
- The *jejunum*, the middle portion, is the shortest segment.
- The *ileum* is the most inferior portion.

Intestinal wall

The intestinal wall has structural features that significantly increase its absorptive surface area. These features include *plicae circulares* — circular folds of the intestinal mucosa, or mucous membrane lining.

Free villi

Villi and microvilli are also intestinal wall features. *Villi* are fingerlike projections on the mucosa. *Microvilli* are tiny cytoplasmic projections on the surface of epithelial cells.

Other structures

The small intestine also contains intestinal crypts, Peyer's patches, and Brunner's glands.

Groovy crypts

Intestinal crypts are simple glands lodged in the grooves separating villi.

Patchwork

Peyer's patches are collections of lymphatic tissue within the submucosa.

(Text continues on page 203.)

Incredibly Easy miniguide: The brain

The illustrations below show the four lobes of the brain and structures of the limbic system.

Each of my cerebral hemispheres is divided into four lobes, and each lobe is named for the cranial bones that lay over it.

Lobes of the brain

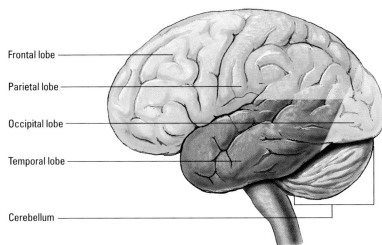

Frontal lobe

Parietal lobe

Occipital lobe

Temporal lobe

Cerebellum

The *limbic system* is a primitive brain area deep within the temporal lobe that initiates basic drives, such as hunger, aggression, and emotional and sexual arousal. It also screens all sensory messages traveling to the cerebral cortex.

Limbic system

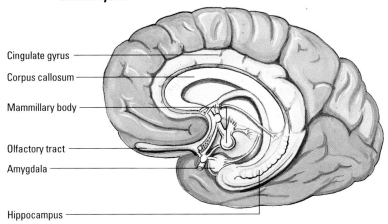

Cingulate gyrus

Corpus callosum

Mammillary body

Olfactory tract

Amygdala

Hippocampus

Incredibly Easy miniguide: The skeletal system

These illustrations show anterior and posterior views of some of the major bones.

I'm important. After all, bones protect internal tissues and organs, stabilize and support the body, allow motion, and provide a surface for muscle, ligament, and tendon attachment.

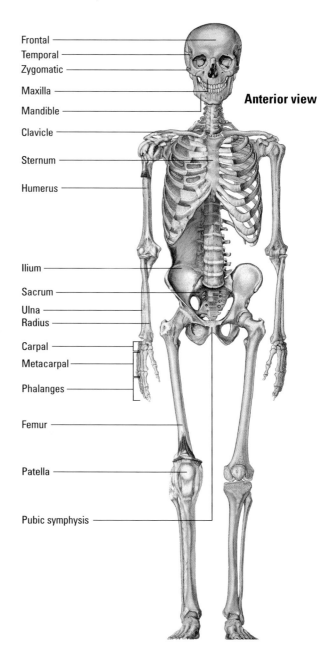

Frontal
Temporal
Zygomatic
Maxilla
Mandible
Clavicle
Sternum
Humerus

Ilium
Sacrum
Ulna
Radius
Carpal
Metacarpal
Phalanges
Femur
Patella
Pubic symphysis

Anterior view

Incredibly Easy miniguide: The skeletal system

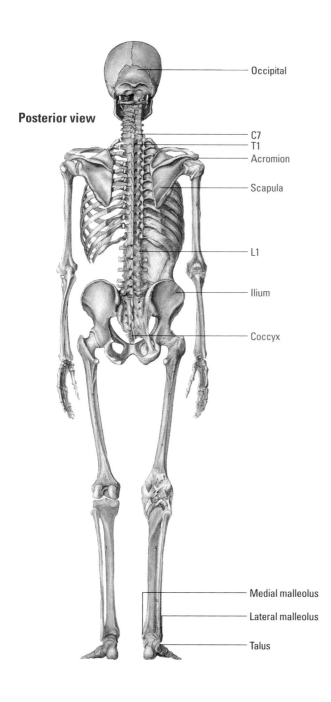

Posterior view

Occipital

C7
T1
Acromion

Scapula

L1

Ilium

Coccyx

Medial malleolus

Lateral malleolus

Talus

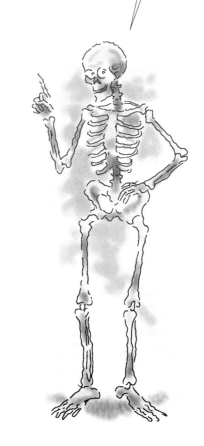

That's not all. Bones also produce red blood cells in the bone marrow and store mineral salts such as calcium.

Incredibly Easy miniguide: The muscular system

These illustrations show some of the major muscles of the body.

Anterior view

Deltoid

Pectoralis major

Rectus abdominis

Biceps brachii

Brachialis

External abdominal oblique

Flexor pollicis longus

Abductor pollicis longus

Pronator quadratus

Flexor retinaculum

Vastus intermedius

Vastus lateralis

Vastus medialis

Patellar ligament

Soleus

Tibialis anterior

Skeletal muscles move body parts or the body as a whole. They're used for both voluntary and reflex movements.

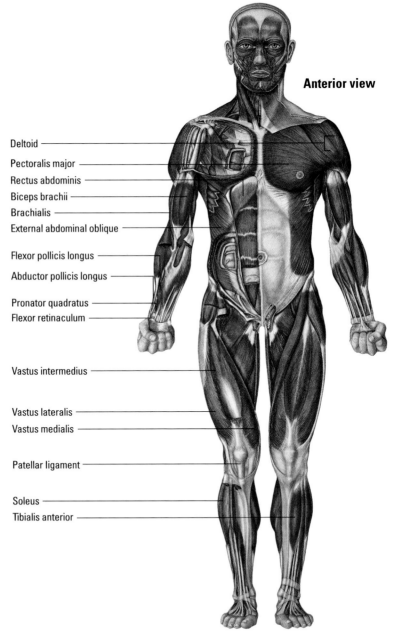

Incredibly Easy miniguide: The muscular system

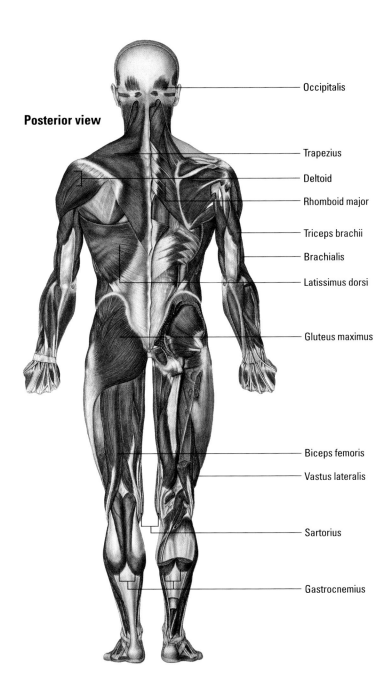

Posterior view

- Occipitalis
- Trapezius
- Deltoid
- Rhomboid major
- Triceps brachii
- Brachialis
- Latissimus dorsi
- Gluteus maximus
- Biceps femoris
- Vastus lateralis
- Sartorius
- Gastrocnemius

Skeletal muscles also maintain posture and generate body heat.

Incredibly Easy miniguide: The heart

The heart is located beneath the sternum in the mediastinum. It rests on the superior surface of the diaphragm and is anterior to the vertebral column.

Anterior view

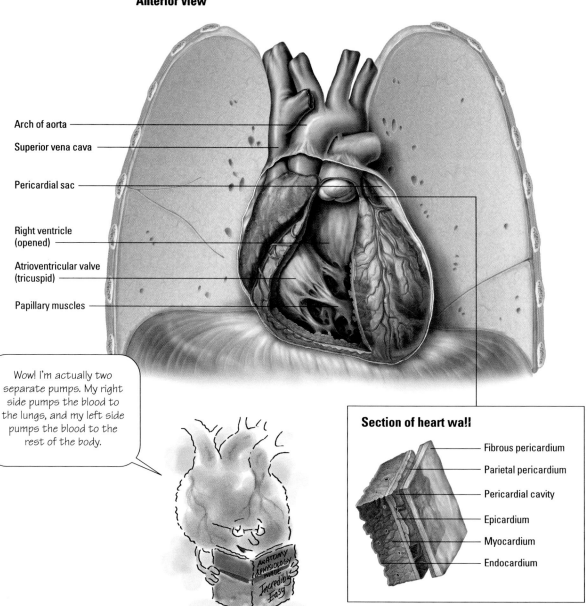

Arch of aorta

Superior vena cava

Pericardial sac

Right ventricle (opened)

Atrioventricular valve (tricuspid)

Papillary muscles

Wow! I'm actually two separate pumps. My right side pumps the blood to the lungs, and my left side pumps the blood to the rest of the body.

Section of heart wall

Fibrous pericardium

Parietal pericardium

Pericardial cavity

Epicardium

Myocardium

Endocardium

Incredibly Easy miniguide: The heart

The myocardium receives blood through four main coronary arteries, as shown below.

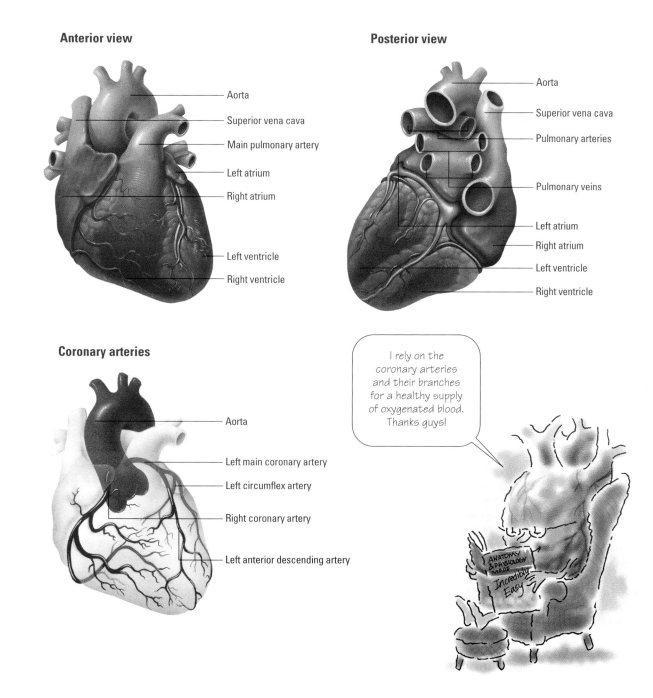

Anterior view

- Aorta
- Superior vena cava
- Main pulmonary artery
- Left atrium
- Right atrium
- Left ventricle
- Right ventricle

Posterior view

- Aorta
- Superior vena cava
- Pulmonary arteries
- Pulmonary veins
- Left atrium
- Right atrium
- Left ventricle
- Right ventricle

Coronary arteries

- Aorta
- Left main coronary artery
- Left circumflex artery
- Right coronary artery
- Left anterior descending artery

I rely on the coronary arteries and their branches for a healthy supply of oxygenated blood. Thanks guys!

ANATOMY & PHYSIOLOGY MADE *Incredibly Easy*

Incredibly Easy miniguide: The respiratory system

The pathways through which air travels to reach the lungs include the nasal and oral cavities, oropharynx, trachea, and bronchi. Air is warmed, filtered, moistened, and delivered to and from the gas exchange area of the lungs.

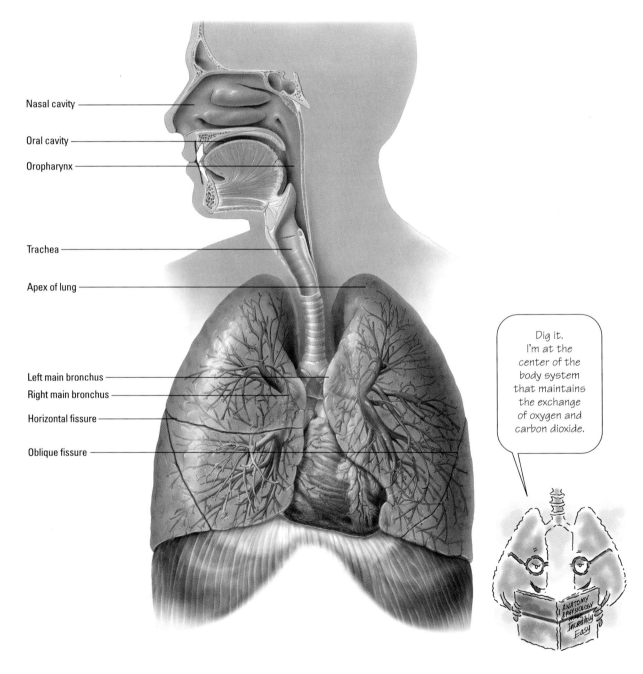

Nasal cavity

Oral cavity

Oropharynx

Trachea

Apex of lung

Left main bronchus

Right main bronchus

Horizontal fissure

Oblique fissure

Dig it. I'm at the center of the body system that maintains the exchange of oxygen and carbon dioxide.

ANATOMY & PHYSIOLOGY made Incredibly Easy

Incredibly Easy miniguide: The respiratory system

The respiratory unit consists of the respiratory bronchiole, alveolar duct and sac, and alveoli. Gas exchange occurs rapidly in the alveoli, in which oxygen from inhaled air diffuses into the blood and carbon dioxide diffuses from the blood into exhaled air.

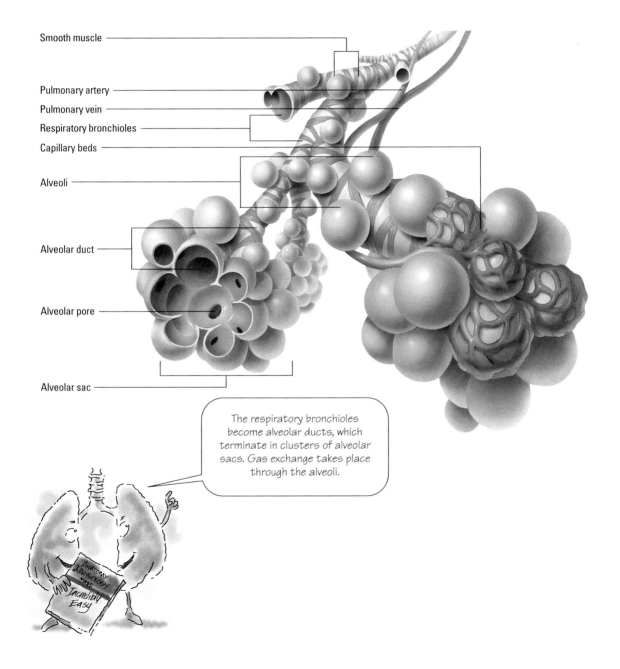

Smooth muscle

Pulmonary artery

Pulmonary vein

Respiratory bronchioles

Capillary beds

Alveoli

Alveolar duct

Alveolar pore

Alveolar sac

The respiratory bronchioles become alveolar ducts, which terminate in clusters of alveolar sacs. Gas exchange takes place through the alveoli.

Incredibly Easy miniguide: The spinal nerves

There are 31 pairs of spinal nerves. After leaving the spinal cord, many nerves join together to form networks called plexuses, as this illustration shows.

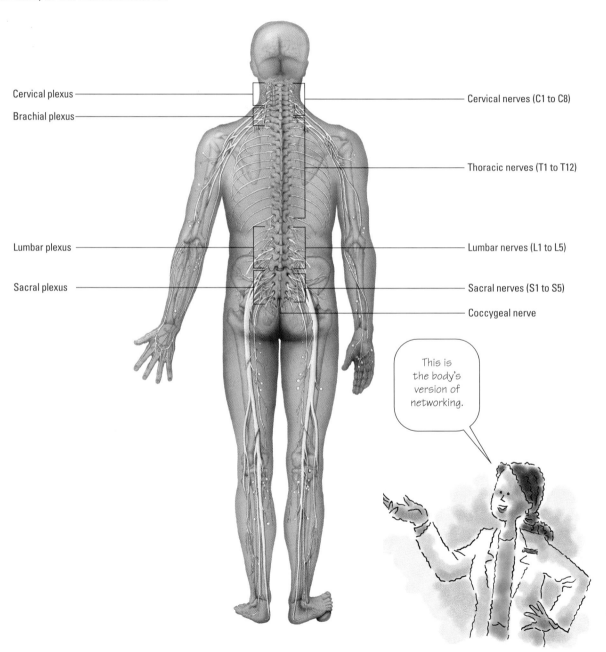

Cervical plexus

Brachial plexus

Lumbar plexus

Sacral plexus

Cervical nerves (C1 to C8)

Thoracic nerves (T1 to T12)

Lumbar nerves (L1 to L5)

Sacral nerves (S1 to S5)

Coccygeal nerve

This is the body's version of networking.

Incredibly Easy miniguide: The endocrine system

The endocrine system regulates and integrates the body's metabolic activities. This illustration shows the locations of the major endocrine glands.

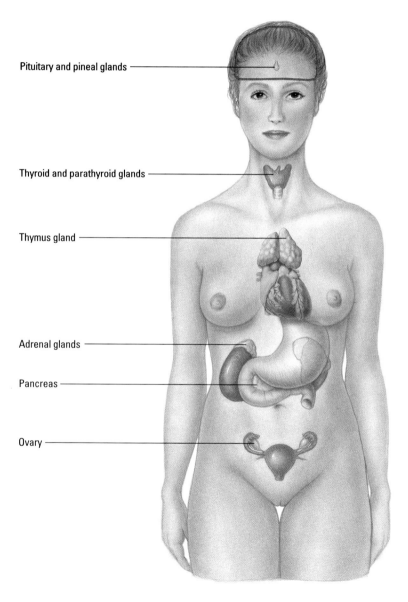

Pituitary and pineal glands

Thyroid and parathyroid glands

Thymus gland

Adrenal glands

Pancreas

Ovary

Along with the nervous system, the endocrine system regulates and integrates the body's metabolic activities.

Incredibly Easy miniguide: The gastrointestinal system

The GI system includes the alimentary canal (pharynx, esophagus, stomach, and small and large intestines) and the accessory organs (salivary glands, liver, biliary system, and pancreas).

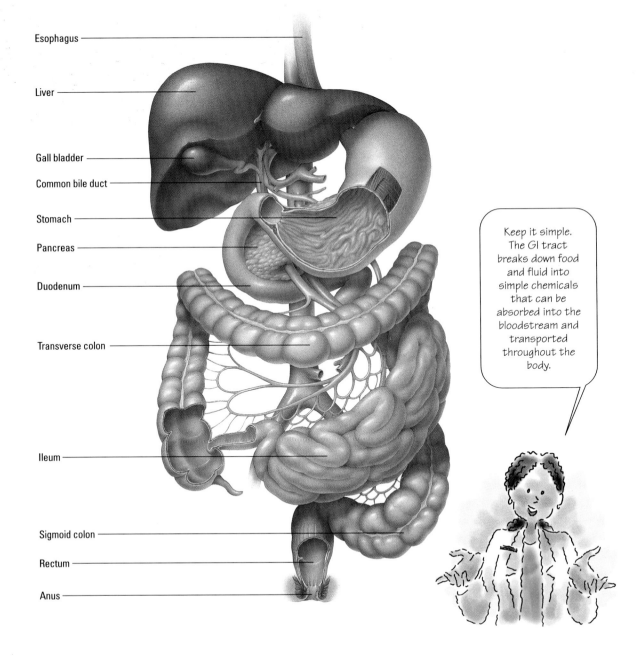

Esophagus

Liver

Gall bladder

Common bile duct

Stomach

Pancreas

Duodenum

Transverse colon

Ileum

Sigmoid colon

Rectum

Anus

Keep it simple. The GI tract breaks down food and fluid into simple chemicals that can be absorbed into the bloodstream and transported throughout the body.

Incredibly Easy miniguide: The urinary system

The urinary system consists of two kidneys, two ureters, the bladder, and the urethra.

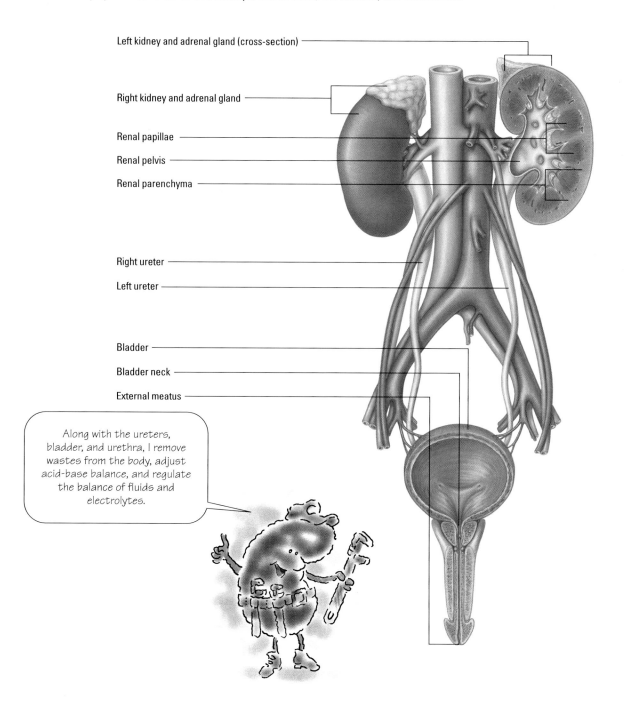

Left kidney and adrenal gland (cross-section)

Right kidney and adrenal gland

Renal papillae

Renal pelvis

Renal parenchyma

Right ureter

Left ureter

Bladder

Bladder neck

External meatus

Along with the ureters, bladder, and urethra, I remove wastes from the body, adjust acid-base balance, and regulate the balance of fluids and electrolytes.

Incredibly Easy miniguide: The male reproductive system

The male reproductive system consists of the penis, scrotum, testes, duct system, and accessory structures (prostate gland, seminal vesicles, bulbourethral glands, and urethral glands).

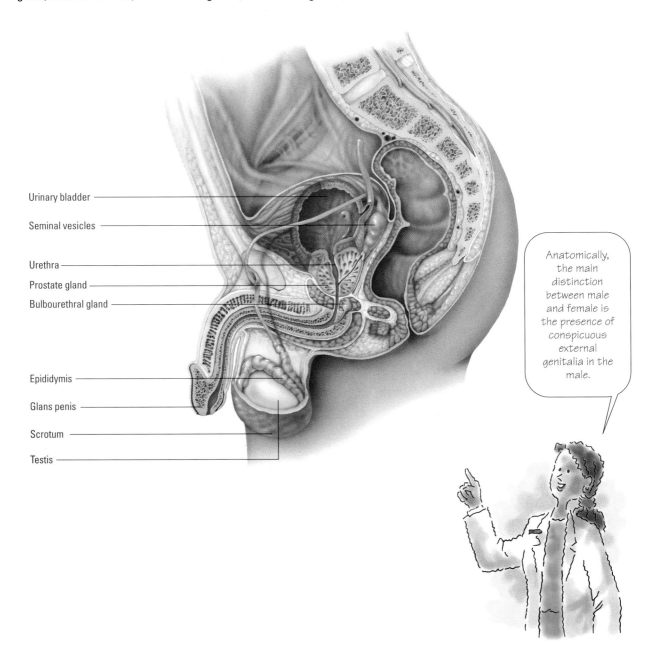

Urinary bladder

Seminal vesicles

Urethra

Prostate gland

Bulbourethral gland

Epididymis

Glans penis

Scrotum

Testis

Anatomically, the main distinction between male and female is the presence of conspicuous external genitalia in the male.

Incredibly Easy miniguide: The female reproductive system

The female reproductive system includes the vagina, uterus, fallopian tubes, ovaries, and external genitalia (not shown).

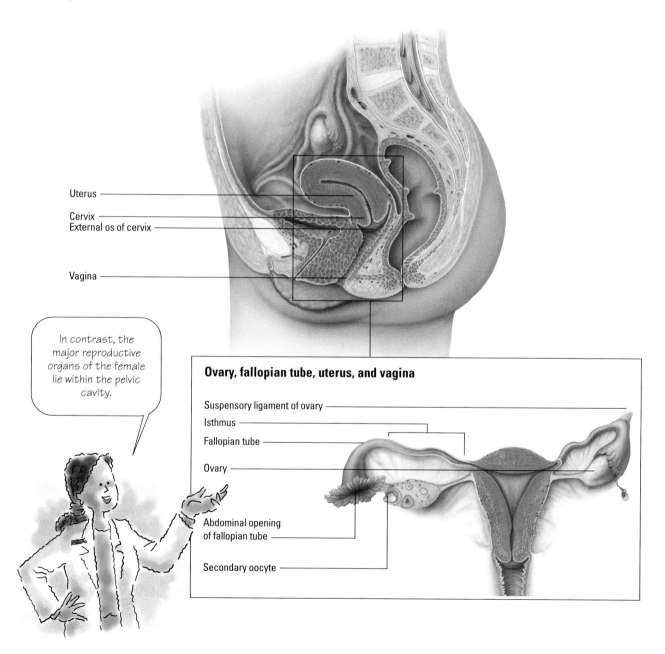

Uterus

Cervix

External os of cervix

Vagina

In contrast, the major reproductive organs of the female lie within the pelvic cavity.

Ovary, fallopian tube, uterus, and vagina

Suspensory ligament of ovary

Isthmus

Fallopian tube

Ovary

Abdominal opening of fallopian tube

Secondary oocyte

Incredibly Easy miniguide: The skin

The skin is composed of two fused layers—the epidermis and dermis.

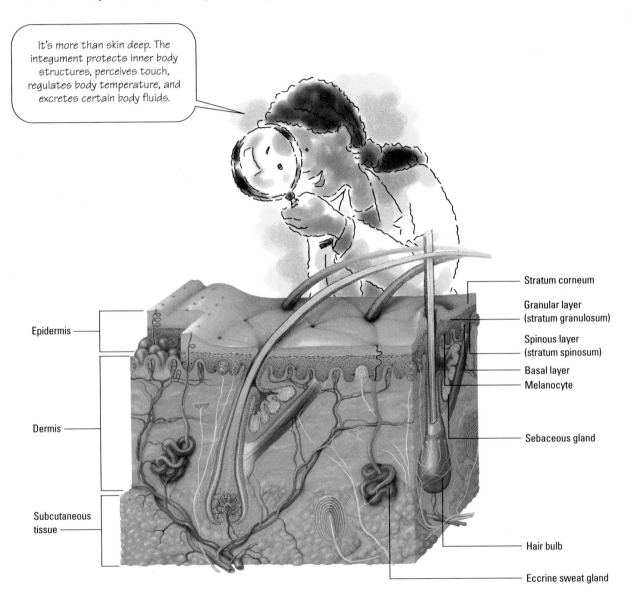

It's more than skin deep. The integument protects inner body structures, perceives touch, regulates body temperature, and excretes certain body fluids.

Epidermis

Dermis

Subcutaneous tissue

Stratum corneum

Granular layer (stratum granulosum)

Spinous layer (stratum spinosum)

Basal layer

Melanocyte

Sebaceous gland

Hair bulb

Eccrine sweat gland

Just another secreting gland

Brunner's glands secrete mucus.

Functions

The small intestine functions by:
- completing food digestion
- absorbing food molecules through its wall into the circulatory system, which then delivers them to body cells
- secreting hormones that help control secretion of bile, pancreatic juice, and intestinal juice.

> Completing digestion, absorbing food, and controlling the secretion of bile...the small intestine has a lot of work to do!

Large intestine

The *large intestine* extends from the ileocecal valve (the valve between the ileum of the small intestine and the first segment of the large intestine) to the anus. The large intestine has six segments:

The *cecum*, a saclike structure, makes up the first few inches.

The *ascending colon* rises on the right posterior abdominal wall, then turns sharply under the liver at the hepatic flexure.

The *transverse colon* is situated above the small intestine, passing horizontally across the abdomen and below the liver, stomach, and spleen. At the left colic flexure, it turns downward.

The *descending colon* starts near the spleen and extends down the left side of the abdomen into the pelvic cavity.

The *sigmoid colon* descends through the pelvic cavity, where it becomes the rectum.

The *rectum*, the last few inches of the large intestine, terminates at the anus.

Functions

The large intestine:
- absorbs water
- secretes mucus
- eliminates digestive wastes.

GI tract wall structures

The wall of the GI tract consists of several layers. These layers are the *mucosa, submucosa, tunica muscularis,* and *visceral peritoneum.*

Past the lips, past the gums...look out, mucosa, here it comes!

Mucosa

Mucosa, the innermost layer, also called the *tunica mucosa,* consists of epithelial and surface cells and loose connective tissue. *Villi,* fingerlike projections of the mucosa, secrete gastric and protective juices and absorb nutrients.

Submucosa

Submucosa, also called the *tunica submucosa,* encircles the mucosa. It's composed of loose connective tissue, blood and lymphatic vessels, and a nerve network (*submucosal plexus* or *Meissner's plexus*).

Tunica muscularis

Tunica muscularis, which lies around the submucosa, is composed of skeletal muscle in the mouth, pharynx, and upper esophagus.

Elsewhere in the tract

Elsewhere in the tract, the tunica muscularis is made up of longitudinal and circular smooth muscle fibers. During peristalsis, longitudinal fibers shorten the lumen length and circular fibers reduce the lumen diameter. At points along the tract, circular fibers thicken to form sphincters.

Pucker pouches

In the large intestine, these fibers gather into three narrow bands (*taeniae coli*) down the middle of the colon and pucker the intestine into characteristic pouches (*haustra*).

The nerve network

Between the two muscle layers lies another nerve network — the *myenteric plexus,* also known as *Auerbach's plexus.* The stomach wall contains a third muscle layer made up of oblique fibers. (See *Features of the GI tract wall.*)

Zoom in

Features of the GI tract wall

Several layers — the tunica mucosa, tunica submucosa, and tunica adventitia — form the wall of the GI tract. This illustration depicts the cellular anatomy of the wall, including special features (such as the villi), the peritoneum, the muscles, and a nerve network.

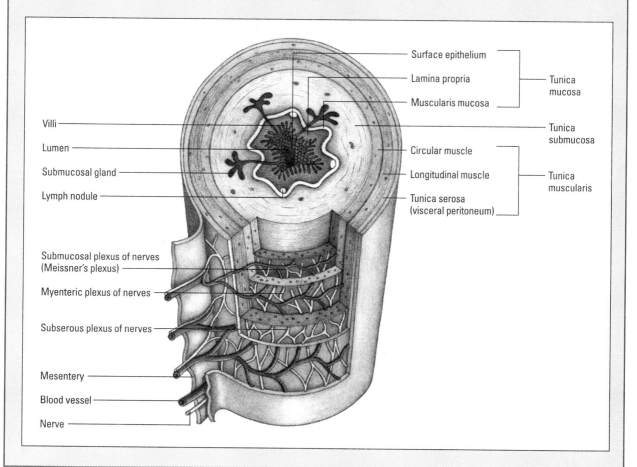

Visceral peritoneum

Visceral peritoneum is the GI tract's outer covering. It covers most of the abdominal organs and lies next to an identical layer, the *parietal peritoneum*, which lines the abdominal cavity.

A double-layered fold

The visceral peritoneum becomes a double-layered fold around the blood vessels, nerves, and lymphatics. It attaches the jejunum and ileum to the posterior abdominal wall to prevent twisting. A similar fold attaches the transverse colon to the posterior abdominal wall.

Named by location

The visceral peritoneum has many names. In the esophagus and rectum, it's called the *tunica adventitia;* elsewhere in the GI tract, it's called the *tunica serosa.*

GI tract innervation

Distention of the submucosal plexus, or Meissner's plexus, stimulates transmission of nerve signals to the smooth muscle, which initiates peristalsis and mixing contractions.

Parasympathetic stimulation

Parasympathetic stimulation of the vagus nerve (for most of the intestines) and the sacral spinal nerves (for the descending colon and rectum) increases gut and sphincter tone. It also increases the frequency, strength, and velocity of smooth muscle contractions as well as motor and secretory activities.

Sympathetic stimulation

Sympathetic stimulation, by way of the spinal nerves from levels T6 to L2, reduces peristalsis and inhibits GI activity.

So many ways of saying one thing! The visceral peritoneum is known as the tunica adventitia in one part of the body and the tunica serosa in another.

Accessory organs of digestion

Accessory organs — the liver, biliary duct system, and pancreas — contribute hormones, enzymes, and bile, which are vital to digestion.

Liver

The body's largest gland, the 3-lb (1.4-kg), highly vascular liver is enclosed in a fibrous capsule in the right upper quadrant of the abdomen.

The *lesser omentum*, a fold of peritoneum, covers most of the liver and anchors the liver to the lesser curvature of the stomach. The *hepatic artery* and *hepatic portal vein* as well as the common bile duct and hepatic veins pass through the lesser omentum.

I'm the largest gland in the body.

Lobes and lobules

The liver consists of four lobes:

 left lobe

 right lobe

 caudate lobe (behind the right lobe)

 quadrate lobe (behind the left lobe).

Functioning features

The liver's functional unit, the *lobule*, consists of a plate of hepatic cells, or *hepatocytes*, that encircle a central vein and radiate outward. Separating the hepatocyte plates from each other are *sinusoids*, the liver's capillary system. Reticuloendothelial macrophages (Kupffer's cells) that lines the sinusoids remove bacteria and toxins that have entered the blood through the intestinal capillaries. (See *Looking at a liver lobule*, page 208.)

Blood flow

The sinusoids carry oxygenated blood from the hepatic artery and nutrient-rich blood from the portal vein. Unoxygenated blood leaves through the central vein and flows through hepatic veins to the inferior vena cava.

Zoom in

Looking at a liver lobule

The liver's functional unit is called a lobule. It consists of a plate of hepatic cells, or *hepatocytes,* that encircle a central vein and radiate outward. Separating the hepatocyte plates from each other are *sinusoids,* which are the liver's capillary system. Sinusoids carry oxygenated blood from the hepatic artery and nutrient-rich blood from the portal vein.

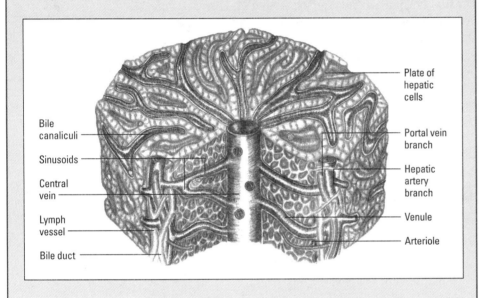

Bile canaliculi

Sinusoids

Central vein

Lymph vessel

Bile duct

Plate of hepatic cells

Portal vein branch

Hepatic artery branch

Venule

Arteriole

Ducts

Think of ducts as a subway system transporting bile through the GI tract. *Bile* is a greenish liquid composed of water, cholesterol, bile salts, and phospholipids. It exits through bile ducts (canaliculi) that merge into the right and left hepatic ducts to form the common hepatic duct. This duct joins the cystic duct from the gallbladder to form the common bile duct leading to the duodenum.

Job description

The liver:
• plays an important role in carbohydrate metabolism
• detoxifies various endogenous and exogenous toxins in plasma

My role in carbohydrate metabolism is very important. I also detoxify toxins in plasma.

- synthesizes plasma proteins, nonessential amino acids, and vitamin A
- stores essential nutrients, such as vitamins K, D, and B_{12} and iron
- removes ammonia from body fluids, converting it to urea for excretion in urine
- helps regulate blood glucose levels
- secretes bile.

Function of bile

Bile has several functions, including:
- emulsifying (breaking down) fat
- promoting intestinal absorption of fatty acids, cholesterol, and other lipids.

When bile salts are MIA

When bile salts are absent from the intestinal tract, lipids are excreted and fat-soluble vitamins are poorly absorbed.

Report on bile production

The liver recycles about 80% of bile salts into bile, combining them with bile pigments (biliverdin and bilirubin, the waste products of red blood cell breakdown) and cholesterol. The liver continuously secretes this alkaline bile. Bile production may increase from stimulation of the vagus nerve, release of the hormone secretin, increased blood flow in the liver, and the presence of fat in the intestine. (See *GI hormones: Production and function*, page 210.)

Bile helps prevent jaundice by aiding excretion of conjugated bilirubin from the liver.

Gallbladder

The *gallbladder* is a pear-shaped organ joined to the ventral surface of the liver by the cystic duct. It's covered with visceral peritoneum.

Job description

The gallbladder stores and concentrates bile produced by the liver. It also releases bile into the common bile duct for delivery to the duodenum in response to the contraction and relaxation of Oddi's sphincter.

Now I get it!

GI hormones: Production and function

When stimulated, GI structures secrete four hormones. Each hormone plays a different part in digestion.

Hormone and production site	Stimulating factor or agent	Function
Gastrin Produced in pyloric antrum and duodenal mucosa	• Pyloric antrum distention • Vagal stimulation • Protein digestion products • Alcohol	Stimulates gastric secretion and motility
Gastric inhibitory peptides Produced in duodenal and jejunal mucosa	• Gastric acid • Fats • Fat digestion products	Inhibits gastric secretion and motility
Secretin Produced in duodenal and jejunal mucosa	• Gastric acid • Fat digestion products • Protein digestion products	Stimulates secretion of bile and alkaline pancreatic fluid
Cholecystokinin Produced in duodenal and jejunal mucosa	• Fat digestion products • Protein digestion products	Stimulates gallbladder contraction and secretion of enzyme-rich pancreatic fluid

Pancreas

The *pancreas* is a somewhat flat organ that lies behind the stomach. Its head and neck extend into the curve of the duodenum and its tail lies against the spleen. (See *Gallbladder and pancreas.*) It performs both exocrine and endocrine functions.

Exocrine function

The pancreas's exocrine function involves scattered cells that secrete more than 1,000 ml of digestive enzymes every day. Lobules and lobes of the clusters (*acini*) of enzyme-producing cells release their secretions into ducts that merge into the pancreatic duct.

Gallbladder and pancreas

Together, the gallbladder and pancreas constitute the biliary tract. The illustration below shows the parts of the biliary tract.

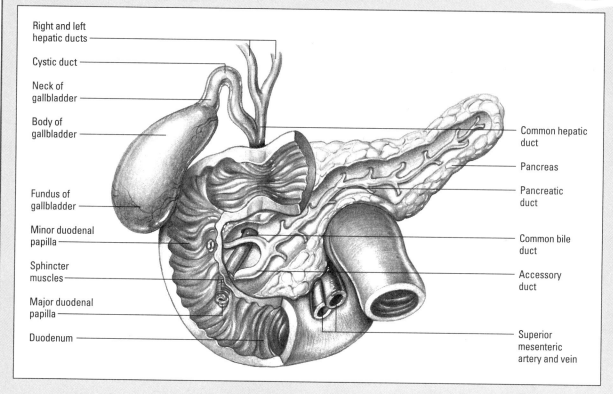

Right and left hepatic ducts

Cystic duct

Neck of gallbladder

Body of gallbladder

Fundus of gallbladder

Minor duodenal papilla

Sphincter muscles

Major duodenal papilla

Duodenum

Common hepatic duct

Pancreas

Pancreatic duct

Common bile duct

Accessory duct

Superior mesenteric artery and vein

Endocrine function

The endocrine function of the pancreas involves the islets of Langerhans, located between the acinar cells.

Alpha and beta cells

Over 1 million of these islets house two cell types: beta and alpha. Beta cells secrete *insulin* to promote carbohydrate metabolism; alpha cells secrete *glucagon*, a hormone that stimulates glycogenolysis in the liver. Both hormones flow directly into the blood. Their release is stimulated by blood glucose levels.

Memory jogger

To remember the difference between exocrine and endocrine, just remember exocrine refers to **ex**ternal, so endocrine refers to internal.

Pancreatic duct

Running the length of the pancreas, the *pancreatic duct* joins the bile duct from the gallbladder before entering the duodenum. Vagal stimulation and release of the hormones secretin and cholecystokinin control the rate and amount of pancreatic secretion.

Digestion and elimination

Chewing, salivation, and swallowing all help begin the process of digestion.

Digestion starts in the oral cavity, where chewing (*mastication*), salivation (the beginning of starch digestion), and swallowing (*deglutition*) all take place.

When a person swallows, the hypopharyngeal sphincter in the upper esophagus relaxes, allowing food to enter the esophagus. (See *What happens in swallowing.*)

Journey of the food bolus

In the esophagus, the glossopharyngeal nerve activates peristalsis, which moves the food down toward the stomach.

As food passes through the esophagus, glands in the esophageal mucosal layer secrete mucus, which lubricates the bolus and protects the mucosal membrane from damage caused by poorly chewed foods.

Cephalic phase of digestion

By the time the food bolus is traveling toward the stomach, the *cephalic phase* of digestion has already begun. In this phase, the stomach secretes digestive juices (hydrochloric acid and pepsin).

Gastric phase of digestion

When food enters the stomach through the cardiac sphincter, the stomach wall stretches, initiating the gastric phase of digestion. In this phase, distention of the stomach wall stimulates the stomach to release *gastrin*.

Now I get it!

What happens in swallowing

Before peristalsis can begin, the neural pattern that initiates swallowing must occur. This process is described here and illustrated below:

• Food pushed to the back of the mouth stimulates swallowing receptor areas that surround the pharyngeal opening.

• These receptor areas transmit impulses to the brain by way of the sensory portions of the trigeminal and glossopharyngeal nerves.

• Then the brain's swallowing center relays motor impulses to the esophagus by way of the trigeminal, glossopharyngeal, vagus, and hypoglossal nerves, causing swallowing to occur.

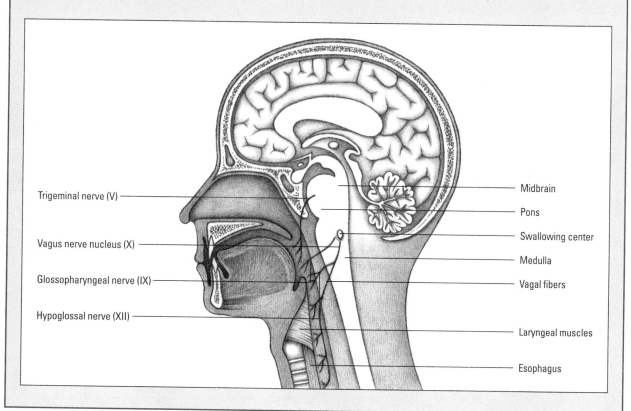

Gastrin

Gastrin stimulates the stomach's motor functions and secretion of gastric juice by the gastric glands. Highly acidic (pH of 0.9 to 1.5), these digestive secretions consist mainly of pepsin, hydrochloric acid, intrinsic factor,

Now I get it!

Sites and mechanisms of gastric secretion

The body of the stomach lies between the lower esophageal, or cardiac, sphincter and the pyloric sphincter. Between these sphincters lie the fundus, body, antrum, and pylorus. These areas have a rich variety of mucosal cells that help the stomach carry out its tasks.

Glands and gastric secretions

Cardiac glands, pyloric glands, and gastric glands secrete 2 to 3 L of gastric juice daily through the stomach's gastric pits.

- Both the *cardiac gland* (near the lower esophageal sphincter [LES]) and the *pyloric gland* (near the pylorus) secrete thin mucus.
- The *gastric gland* (in the body and fundus) secretes hydrochloric acid (HCl), pepsinogen, intrinsic factor, and mucus.

Protection from self-digestion

Specialized cells line the gastric glands, gastric pits, and surface epithelium. Mucous cells in the necks of the gastric glands produce thin mucus. Mucous cells in the surface epithelium produce an alkaline mucus. Both substances lubricate food and protect the stomach from self-digestion by corrosive enzymes.

Other secretions

Argentaffin cells produce gastrin, which stimulates gastric secretion and motility. *Chief cells* produce pepsinogen, which breaks proteins down into polypeptides. Large parietal cells scattered throughout the fundus secrete HCl and intrinsic factor. HCl degrades pepsinogen, maintains acid environment, and inhibits excess bacteria growth. Intrinsic factor promotes vitamin B_{12} absorption in the small intestine.

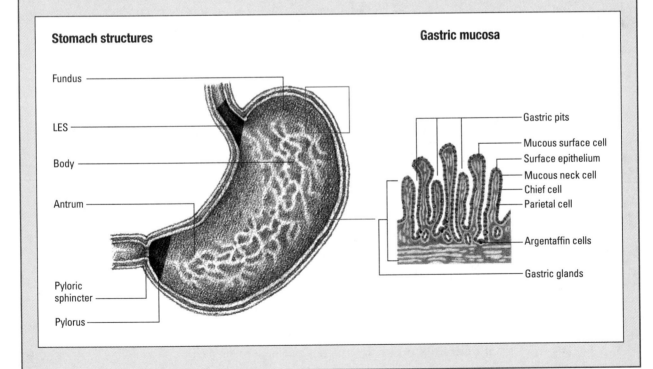

Stomach structures

Fundus
LES
Body
Antrum
Pyloric sphincter
Pylorus

Gastric mucosa

Gastric pits
Mucous surface cell
Surface epithelium
Mucous neck cell
Chief cell
Parietal cell
Argentaffin cells
Gastric glands

and proteolytic enzymes. (See *Sites and mechanisms of gastric secretion.*)

Intestinal phase of digestion

Normally, except for alcohol, little food absorption occurs in the stomach. Peristaltic contractions churn the food into tiny particles and mix it with gastric juices, forming chyme.

Next, stronger peristaltic waves move the chyme into the antrum, where it backs up against the pyloric sphincter before being released into the duodenum, triggering the intestinal phase of digestion.

Stomach emptying

The rate of stomach emptying depends on several factors, including gastrin release, neural signals generated when the stomach wall distends, and the *enterogastric reflex*. In this reaction, the duodenum releases secretin and gastric-inhibiting peptide, and the jejunum secretes cholecystokinin — all of which act to decrease gastric motility.

What happens in the small intestine...

The small intestine performs most of the work of digestion and absorption. (See *Small intestine: How form affects absorption*, pages 216 and 217.)

In the small intestine, intestinal contractions and various digestive secretions break down carbohydrates, proteins, and fats — actions that enable the intestinal mucosa to absorb these nutrients into the bloodstream (along with water and electrolytes). These nutrients are then available for use by the body.

...and what happens in the large intestine

By the time chyme passes through the small intestine and enters the ascending colon of the large intestine, it has been reduced to mostly indigestible substances.

On the road with the food bolus

The food bolus begins its journey through the large intestine where the ileum and cecum join with the ileocecal pouch. Then the bolus moves up the ascending colon, past the right abdominal cavity to the liver's lower bor-

> Digestive juices are secreted in response to smelling, tasting, chewing, or thinking about food.

> Alcohol is one of the few substances that is digested primarily in the stomach.

(Text continues on page 218.)

Now I get it!

Small intestine: How form affects absorption

Nearly all digestion and absorption takes place in the 20′ (6.1 m) of small intestine. The structure of the small intestine, as shown below, is key to digestion and absorption.

Specialized mucosa

Multiple projections of the intestinal mucosa increase the surface area for absorption several hundredfold, as shown in the enlarged views below.

Circular projections *(Kerckring's folds)* are covered by villi. Each villus contains a lymphatic vessel *(lacteal),* a venule, capillaries, an arteriole, nerve fibers, and smooth muscle.

Each villus is densely fringed with about 2,000 microvilli making it resemble a fine brush. The villi are lined with columnar epithelial cells, which dip into the lamina propria between the villi to form intestinal glands *(crypts of Lieberkühn).*

Types of epithelial cells

The type of epithelial cell dictates its function. Mucus-secreting goblet cells are found on and between the villi on the crypt mucosa. In the proximal duodenum, specialized Brunner's glands

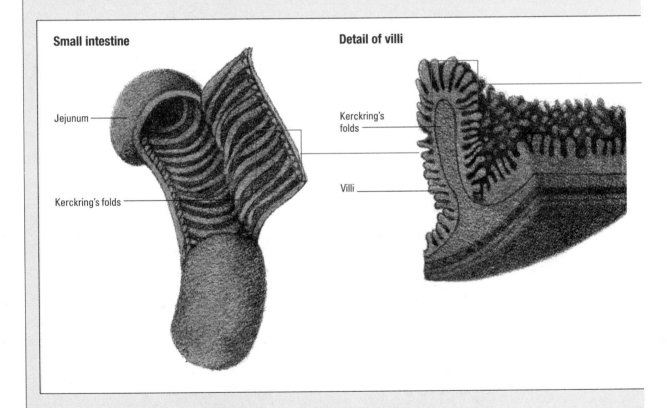

Small intestine

Jejunum

Kerckring's folds

Detail of villi

Kerckring's folds

Villi

also secrete large amounts of mucus to lubricate and protect the duodenum from potentially corrosive acidic chyme and gastric juices.

Duodenal *argentaffin cells* produce the hormones secretin and cholecystokinin. *Undifferentiated cells* deep within the intestinal glands replace the epithelium. *Absorptive cells* consist of large numbers of tightly packed microvilli over a plasma membrane that contains transport mechanisms for absorption and produces enzymes for the final step in digestion.

Intestinal glands

The intestinal glands primarily secrete a watery fluid that bathes the villi with chyme particles. Fluid production results from local irritation of nerve cells and possibly from hormonal stimulation by secretin and cholecystokinin. The microvillous brush border secretes various hormones and digestive enzymes that catalyze final nutrient breakdown.

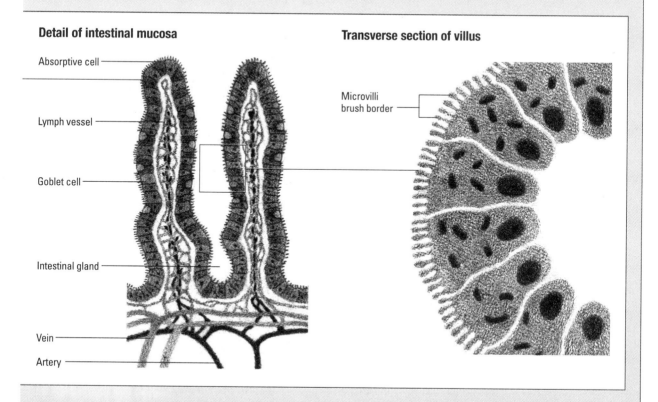

Detail of intestinal mucosa

Absorptive cell

Lymph vessel

Goblet cell

Intestinal gland

Vein

Artery

Transverse section of villus

Microvilli brush border

der. It crosses horizontally below the liver and stomach by way of the transverse colon and descends the left abdominal cavity to the iliac fossa through the descending colon.

More travels with the food bolus

From there, the bolus travels through the sigmoid colon to the lower midline of the abdominal cavity, then to the rectum, and finally to the anal canal. The anus opens to the exterior through two sphincters. The *internal sphincter* contains thick, circular smooth muscle under autonomic control; the *external sphincter* contains skeletal muscle under voluntary control.

Role in absorption

The large intestine produces no hormones or digestive enzymes; it continues the absorptive process. Through blood and lymph vessels in the submucosa, the proximal half of the large intestine absorbs all but about 100 ml of the remaining water in the colon. It also absorbs large amounts of both sodium and chloride.

Bacterial action

The large intestine harbors the bacteria *Escherichia coli*, *Enterobacter aerogenes*, *Clostridium perfringens*, and *Lactobacillus bifidus*. All of these bacteria help synthesize vitamin K and break down cellulose into a usable carbohydrate. Bacterial action also produces *flatus*, which helps propel stool toward the rectum.

Protection from bacterial action

In addition, the mucosa of the large intestine produces *alkaline secretions* from tubular glands composed of goblet cells. This alkaline mucus lubricates the intestinal walls as food pushes through, protecting the mucosa from acidic bacterial action.

Mass movement

In the lower colon, long and relatively sluggish contractions cause propulsive waves, or *mass movements*. Normally occurring several times per day, these movements propel intestinal contents into the rectum and produce the urge to defecate.

Defecation normally results from the *defecation reflex*, a sensory and parasympathetic nerve-mediated response, along with the voluntary relaxation of the external anal sphincter.

Quick quiz

1. Which of the following is *not* a function of the stomach?

    A.   Acting as a temporary storage area

    B.   Mixing food with gastric secretions

    C.   Completing food digestion

Answer: C. The small intestine, not the stomach, completes digestion.

2. Which of the following is *not* a function of the liver?

    A.   Detoxification of endogenous and exogenous toxins in plasma

    B.   Storage of vitamins A, C, and E

    C.   Storage of vitamins K, D, and B_{12}

Answer: B. The liver doesn't store vitamins A, C, or E. It only synthesizes vitamin A.

3. Which GI hormone stimulates gastric secretion and motility?

    A.   Gastrin

    B.   Gastric inhibitory peptides

    C.   Secretin

Answer: A. Gastrin is produced in the pyloric antrum and duodenal end mucosa and stimulates gastric secretion and motility.

4. In which phase of digestion does the stomach secrete the digestive juices hydrochloric acid and pepsin?

    A.   Cephalic

    B.   Gastric

    C.   Intestinal

Answer: A. By the time food is traveling toward the stomach, the cephalic phase — during which the stomach secretes digestive juices — has begun.

Scoring

☆☆☆   If you answered all four questions correctly, bravo! You've passed through the GI system with the greatest of ease!

☆☆   If you answered two or three questions correctly, super! You've chewed the fat of this system and it's time to move on.

☆   If you answered fewer than two questions correctly, don't worry! It might take a little longer to digest this material, but you've also got five more quick quizzes to go, so keep at it!

Nutrition and metabolism

Just the facts

In this chapter, you'll learn:

♦ role of carbohydrates, proteins, and lipids in nutrition

♦ function of vitamins and minerals in the body

♦ how glucose is turned to energy

♦ role of hormones in metabolism.

Nutrition

Nutrition refers to the taking in, assimilating, and utilizing of nutrients. The crucial nutrients in foods must be broken down into components. Within cells, the products of digestion undergo further chemical reactions.

Metabolism refers to the sum of these chemical reactions. Through metabolism, food substances are transformed into energy or materials that the body can use or store.

Metabolism involves two processes:

☝ *anabolism* — synthesis of simple substances into complex ones

✌ *catabolism* — breakdown of complex substances into simpler ones or into energy.

Food
+
Metabolism =

Energy

Needed for nutrition

The body needs a continual supply of water and various nutrients for growth and repair. Virtually all nutrients come from digested food. The three major types of nutrients required by the body are carbohydrates, proteins, and lipids.

Needed for metabolism

Vitamins are essential for normal metabolism. They contribute to the enzyme reactions that promote the metabolism of carbohydrates, proteins, and lipids.

Minerals are also important. They participate in such essential functions as enzyme metabolism and membrane transfer of essential elements. (See *The food pyramid.*)

Carbohydrates

Carbohydrates are organic compounds composed of carbon, hydrogen, and oxygen; they yield 4 kcal/g when used for energy. Carbohydrates are classified as *monosaccha-*

The energy in nutrients is measured in kilocalories (kcal) — commonly called calories — per gram of nutrient. Adults need approximately 2,000 kcal daily.

The food pyramid

This food pyramid suggests eating a large variety of foods (with emphasis on fruits, vegetables, and starches) for good health. In this diet, most of the fat comes from the milk and cheese and meat, poultry, and eggs groups. Caloric intake ranges from 1,500 calories to 2,800 calories daily.

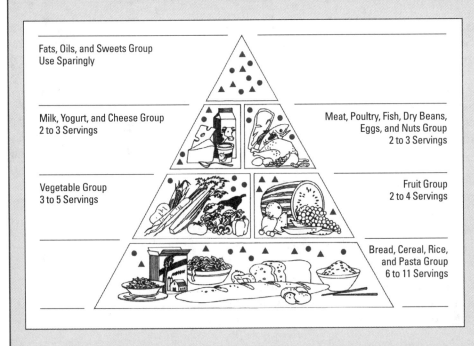

Fats, Oils, and Sweets Group
Use Sparingly

Milk, Yogurt, and Cheese Group
2 to 3 Servings

Meat, Poultry, Fish, Dry Beans, Eggs, and Nuts Group
2 to 3 Servings

Vegetable Group
3 to 5 Servings

Fruit Group
2 to 4 Servings

Bread, Cereal, Rice, and Pasta Group
6 to 11 Servings

rides, disaccharides, and *polysaccharides.* Sugars are carbohydrates and function as the body's primary energy source.

Monosaccharides

Monosaccharides are simple sugars that can't be split into smaller units by hydrolysis. They're subdivided into polyhydroxy aldehydes or ketones based on whether the molecule consists of an aldehyde group or a ketone group.

Monosaccharides are also classified by the number of carbon atoms they contain.

The OH link

An aldehyde contains the characteristic group CHO. The term *polyhydroxy* refers to the linking of carbon atoms to a hydroxyl (OH) group.

CO is the key to ketone

A *ketone,* on the other hand, contains the carbonyl group CO and carbon groups attached to the carbonyl carbon.

Disaccharides

Disaccharides are synthesized from monosaccharides. A disaccharide molecule consists of two monosaccharides minus a water molecule. Examples of disaccharides include:
• **Sucrose,** common table sugar, is also found in some fruits and vegetables, and is a combination of a glucose molecule and a fructose molecule.
• **Lactose,** the sugar found in milk, is a combination of a glucose molecule and a galactose molecule
• **Maltose,** a sugar used in brewing and distilling, is a combination of two glucose molecules.

Polysaccharides

Like disaccharides, *polysaccharides* are synthesized from monosaccharides. A polysaccharide consists of a long chain (*polymer*) of more than 10 monosaccharides linked by glycoside bonds. Glycogen is an example of a polysaccharide. Your body builds glycogen by using excess sugar (monosaccharides) and stores it for future use. Polysaccharides are also ingested and broken down into simple sugars used for fuel. However, fiber is an example of a polysaccharide that can't be broken down into simple sugars. Thus, the body can't derive energy (fuel) from fiber.

Proteins

Proteins are complex nitrogenous organic compounds containing amino acid chains; some also contain sulfur and phosphorus. Proteins are used mainly for growth and repair of body tissues; when used for energy, they yield 4 kcal/g. Some proteins combine with lipids to form *lipoproteins* or with carbohydrates to form *glyco-proteins*.

Amino acids

Amino acids are the building blocks of proteins. Each amino acid contains a carbon atom to which a carboxyl (COOH) group and an amino group are attached.

Making peptide bonds

Amino acids unite by condensation of the COOH group on one amino acid with the amino group of the adjacent amino acid. This reaction releases a water molecule and creates a linkage called a *peptide bond*.

Shape determines function

The sequence and types of amino acids in the chain determine the nature of the protein. Each protein is synthesized on a ribosome as a straight chain. Chemical attractions between the amino acids in various parts of the chain cause the chain to coil or twist into a specific shape. A protein's shape, in turn, determines its function.

Lipids

Lipids are organic compounds that don't dissolve in water but do dissolve in alcohol and other organic solvents. Lipids are a concentrated form of fuel and yield approximately 9 kcal/g when used for energy. The major lipids include fats (the most common lipids), phospholipids, and steroids.

The body uses proteins for growth and repair.

Memory jogger

Remember the three **P's** of proteins:

Peptide = 2 to 10 amino acids

Polypeptide = 10 or more amino acids

Protein = 50 or more amino acids.

Fats

A fat, or *triglyceride,* contains three molecules of fatty acid combined with one molecule of glycerol. A fatty acid is a long-chain compound with an even number of carbon atoms and a terminal COOH group.

The glycerol example

Glycerol, for example, is a three-carbon compound (alcohol) with an OH group attached to each carbon atom. The COOH group on each fatty acid molecule joins to one OH group on the glycerol molecule; this results in the release of a water molecule. Linking of the COOH and OH groups produces an ester linkage.

Phospholipids

Phospholipids are complex lipids that are similar to fat but have a phosphorus- and nitrogen-containing compound that replaces one of the fatty acid molecules. Phospholipids are major structural components of cell membranes.

Steroids

Steroids are complex molecules in which the carbon atoms form four cyclic structures attached to various side chains. They contain no glycerol or fatty acid molecules. Examples of steroids include cholesterol, bile salts, and sex hormones.

The body must obtain adequate vitamins from the diet because it can't manufacture many vitamins itself.

Vitamins and minerals

Vitamins are organic compounds that are needed in small quantities for normal metabolism, growth, and development.

Vitamins are classified as *water-soluble* or *fat-soluble.* Water-soluble vitamins include the B complex and C vitamins; fat-soluble vitamins include vitamins A, D, E, and K. (See *Guide to vitamins and minerals,* pages 226 to 228.)

(Text continues on page 229.)

Guide to vitamins and minerals

Good health requires intake of adequate amounts of vitamins and minerals to meet the body's metabolic needs. A vitamin or mineral excess or deficiency, although rare, can lead to various disorders. The chart below reviews major functions and food sources for vitamins and minerals.

Vitamin or mineral	Major functions	Food sources
Water-soluble vitamins		
• Vitamin C (ascorbic acid)	• Collagen production, digestion, fine bone and tooth formation, iodine conservation, healing, red blood cell formation, infection resistance	• Fresh fruits and vegetables
• Vitamin B_1 (thiamine)	• Appetite stimulation, blood building, carbohydrate metabolism, circulation, digestion, growth, learning ability, muscle tone maintenance	• Meats, fish, poultry, pork, molasses, brewer's yeast, brown rice, nuts, wheat germ, whole and enriched grains
• Vitamin B_2 (riboflavin)	• Antibody and red blood cell formation; energy metabolism; cell respiration; epithelial, eye, and mucosal tissue maintenance	• Meats, fish, poultry, milk, molasses, brewer's yeast, eggs, fruit, green leafy vegetables, nuts, whole grains
• Vitamin B_6 (pyridoxine)	• Antibody formation, digestion, deoxyribonucleic acid and ribonucleic acid synthesis, fat and protein utilization, amino acid metabolism, hemoglobin production	• Meats, poultry, bananas, molasses, brewer's yeast, desiccated liver, fish, green leafy vegetables, peanuts, raisins, walnuts, wheat germ, whole grains
• Folic acid (folacin, pteroylglutamic acid)	• Cell growth and reproduction, hydrochloric acid production, liver function, nucleic acid formation, protein metabolism, red blood cell formation	• Citrus fruits, eggs, green leafy vegetables, milk products, organ meats, seafood, whole grains
• Niacin (nicotinic acid, nicotinamide, niacinamide)	• Circulation, cholesterol level reduction, growth, hydrochloric acid production, metabolism (carbohydrate, protein, fat), sex hormone production	• Eggs, lean meats, milk products, organ meats, peanuts, poultry, seafood, whole grains
• Vitamin B_{12} (cyanocobalamin)	• Blood cell formation, cellular and nutrient metabolism, iron absorption, tissue growth, nerve cell maintenance	• Beef, eggs, fish, milk products, organ meats, pork

Guide to vitamins and minerals *(continued)*

Vitamin or mineral	Major functions	Food sources
Fat-soluble vitamins		
• Vitamin A	• Body tissue repair and maintenance, infection resistance, bone growth, nervous system development, cell membrane metabolism and structure	• Fish, green and yellow fruits and vegetables, milk products
• Vitamin D (calciferol)	• Calcium and phosphorus metabolism (bone formation), myocardial function, nervous system maintenance, normal blood clotting	• Bonemeal, egg yolks, organ meats, butter, cod liver oil, fatty fish
• Vitamin E (tocopherol)	• Aging retardation, anticlotting factor, diuresis, fertility, lung protection (antipollution), male potency, muscle and nerve cell membrane maintenance, myocardial perfusion, serum cholesterol reduction	• Butter, dark green vegetables, eggs, fruits, nuts, organ meats, vegetable oils, wheat germ
• Vitamin K (menadione)	• Liver synthesis of prothrombin and other blood-clotting factors	• Green leafy vegetables, safflower oil, yogurt, liver, molasses
Minerals		
• Calcium	• Blood clotting, bone and tooth formation, cardiac rhythm, cell membrane permeability, muscle growth and contraction, nerve impulse transmission	• Bonemeal, cheese, milk, molasses, yogurt, whole grains, nuts, legumes, leafy vegetables
• Chloride	• Maintenance of fluid, electrolyte, acid-base, and osmotic pressure balance	• Fruits, vegetables, table salt
• Magnesium	• Acid-base balance, metabolism, protein synthesis, muscle relaxation, cellular respiration, nerve impulse transmission	• Green leafy vegetables, nuts, seafood, cocoa, whole grains
• Phosphorus	• Bone and tooth formation, cell growth and repair	• Eggs, fish, grains, meats, poultry, yellow cheese, milk, milk products
• Potassium	• Heartbeat, muscle contraction, nerve impulse transmission, rapid growth, fluid distribution and osmotic pressure balance, acid-base balance	• Seafood, molasses, peaches, peanuts, raisins

(continued)

Guide to vitamins and minerals (continued)

Vitamin or mineral	Major functions	Food sources
Minerals (continued)		
• Sodium	• Cellular fluid level maintenance, muscle contraction, acid-base balance, cell permeability, muscle function, nerve impulse transmission	• Seafood, cheese, milk, salt
• Fluoride (fluorine)	• Bone and tooth formation	• Drinking water
• Iodine	• Energy production, metabolism, physical and mental development	• Kelp, salt (iodized), seafood
• Iron	• Growth (in children), hemoglobin production, stress and disease resistance, cellular respiration, oxygen transport	• Eggs, organ meats, poultry, wheat germ, liver, potatoes, enriched breads and cereals, green vegetables, molasses
• Selenium	• Immune mechanisms, mitochondrial adenosine triphosphate synthesis, cellular protection, fat metabolism	• Seafood, meats, liver, kidneys
• Zinc	• Burn and wound healing, carbohydrate digestion, metabolism (carbohydrate, fat, protein), prostate gland function, reproductive organ growth and development	• Liver, mushrooms, seafood, soybeans, spinach, meat

Minerals

Minerals are inorganic substances that play important roles in:
• enzyme metabolism
• membrane transfer of essential compounds
• regulation of acid-base balance
• osmotic pressure
• muscle contractility
• nerve impulse transmission
• growth.

Minerals are found in bones, hemoglobin, thyroxine, and vitamin B_{12}.

Major and trace minerals

Minerals may be classified as *major minerals* (more than 0.005% of body weight) or *trace minerals* (less than

0.005% of body weight). Major minerals include calcium, chloride, magnesium, phosphorus, potassium, and sodium. Trace minerals include chromium, cobalt, copper, fluorine, iodine, iron, manganese, molybdenum, selenium, and zinc.

Absorption and digestion

> During hydrolysis, a compound unites with water and then splits into simpler compounds.

Nutrients must be digested in the GI tract by enzymes that split large units into smaller ones. In this process, called *hydrolysis*, a compound unites with water and then splits into simpler compounds. The smaller units are then absorbed from the small intestine and transported to the liver through the portal venous system.

Carbohydrate digestion and absorption

Enzymes break down complex carbohydrates. In the oral cavity, *salivary amylase* initiates starch hydrolysis into disaccharides. In the small intestine, *pancreatic amylase* continues this process.

Breaking it down

Disaccharides in the intestinal mucosa hydrolyze disaccharides into monosaccharides. Lactase splits the compound lactose into glucose and galactose, and sucrase hydrolyzes the compound sucrose into glucose and fructose.

Movin' right along

Monosaccharides, such as glucose, fructose, and galactose, are absorbed through the intestinal mucosa and are then transported through the portal venous system to the liver. There, enzymes convert fructose and galactose to glucose.

Ribonucleases and deoxyribonucleases break down nucleotides from deoxyribonucleic acid and ribonucleic acid into pentoses and nitrogen-containing compounds (nitrogen bases). Like glucose, these compounds are absorbed through the intestinal mucosa.

Protein digestion and absorption

Enzymes digest proteins by hydrolyzing the peptide bonds that link the amino acids of the protein chains, which restores water molecules.

Gastric pepsin breaks proteins into:
- polypeptides
- pancreatic trypsin
- chymotrypsin
- carboxypeptidase, which converts polypeptides to peptides.

The breakdown lane

Intestinal mucosal peptidases break down peptides into their constituent amino acids. After being absorbed through the intestinal mucosa by active transport mechanisms, these amino acids travel through the portal venous system to the liver. The liver converts the amino acids not needed for protein synthesis into glucose.

I convert any amino acids not needed for protein synthesis into glucose.

Lipid digestion and absorption

Pancreatic lipase breaks down fats and phospholipids into a mixture of:
- glycerol
- short- and long-chain fatty acids
- monoglycerides.

The portal venous system then carries these substances to the liver. Lipase hydrolyzes the bonds between glycerol and fatty acids — a process that restores the water molecules released when the bonds were formed.

On a short leash...

Glycerol diffuses directly through the mucosa. Short-chain fatty acids diffuse into the intestinal epithelial cells and are carried to the liver via the portal venous system.

...or on a long one

Long-chain fatty acids and monoglycerides in the intestine dissolve in the bile salt micelles and then diffuse into the intestinal epithelial cells. There, lipase breaks down absorbed monoglycerides into glycerol and fatty acids. In the smooth endoplasmic reticulum of the epithelial cells, fatty acids and glycerol recombine to form fats.

Chylomicrons

Along with a small amount of cholesterol and phospholipid, triglycerides are coated with a thin layer of protein to form lipoprotein particles called *chylomicrons*. Chylomicrons collect in the intestinal lacteals (lymphatic vessels) and are carried through lymphatic channels. After entering the circulation through the thoracic duct, they're distributed to body cells.

Stored away for later

In the cells, fats are extracted from the chylomicrons and broken down by enzymes into fatty acids and glycerol. Then they're absorbed and recombined in fat cells, reforming triglycerides for storage and later use.

Carbohydrate metabolism

Carbohydrates are the preferred energy fuel of human cells. Most of the carbohydrates in absorbed food is quickly catabolized for the release of energy.

Glucose to energy

All ingested carbohydrates are converted to glucose, the body's main energy source. Glucose not needed for immediate energy is stored as glycogen or converted to lipids.

Energy from glucose catabolism is generated in three phases — glycolysis, the Krebs cycle (also called the citric acid cycle), and the electron transport system. (See *Tracking the glucose pathway*, page 232.) Glycolysis, which occurs in the cell cytoplasm, doesn't use oxygen. The other two phases, which occur in mitochondria, do use oxygen.

Glycolysis

Glycolysis refers to the process by which enzymes break down the 6-carbon glucose molecule into two 3-carbon molecules of pyruvic acid (pyruvate). Glycolysis yields energy in the form of adenosine triphosphate (ATP).

This whole nutrition and metabolism thing is pretty simple; it's about getting energy to DO THINGS!

Now I get it!

Tracking the glucose pathway

Glucose catabolism generates energy in three phases: glycolysis, Krebs cycle, and the electron-transport chain. The illustration below summarizes the first two phases.

Glycolysis

Glycolysis, the first phase, breaks apart one molecule of glucose to form two molecules of pyruvate, which yields energy in the form of adenosine triphosphate (ATP) and acetyl coenzyme A (CoA).

Krebs cycle

The second phase, the Krebs cycle, continues carbohydrate metabolism. Fragments of acetyl CoA join to oxaloacetic acid to form citric acid. The CoA molecule breaks off from the acetyl group and may form more acetyl CoA molecules. Citric acid is first converted into intermediate compounds and then back into oxalo-acetic acid. The Krebs cycle also liberates carbon dioxide.

Electron-transport chain

In the third phase of glucose catabolism, molecules on the inner mitochondrial membrane attract electrons from hydrogen atoms and carry them through oxidation-reduction reactions in the mitochondria. The hydrogen ions produced in the Krebs cycle then combine with oxygen to form water.

Glucose is the body's main energy source.

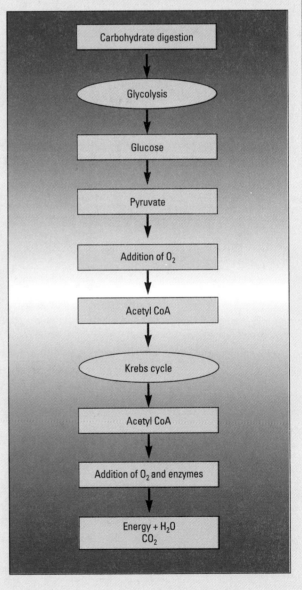

Cruising the glucose pathway

Next, pyruvic acid releases a carbon dioxide (CO_2) molecule and is converted in the mitochondria to a two-carbon acetyl fragment, which combines with coenzyme A (CoA) (a complex organic compound) to form acetyl CoA.

Krebs cycle

The second phase in glucose catabolism is the Krebs cycle. It's the pathway by which a molecule of acetyl CoA is oxidized by enzymes to yield energy.

Liberating carbon dioxide, releasing hydrogen

The two-carbon acetyl fragments of acetyl CoA enter the Krebs cycle by joining to the four-carbon compound oxaloacetic acid to form citric acid, a six-carbon compound. In this process, the CoA molecule detaches from the acetyl group, becoming available to form more acetyl CoA molecules. Enzymes convert citric acid into intermediate compounds and eventually convert it back into oxaloacetic acid. Then the cycle can begin again.

In addition to liberating CO_2 and generating energy, each turn of the Krebs cycle releases hydrogen atoms, which are picked up by the coenzymes nicotinamide adenine dinucleotide (NAD) and flavin adenine dinucleotide (FAD).

> The last step of carbohydrate catabolism is electron transport.

Electron transport system

The electron transport system is the last phase of carbohydrate catabolism. In this phase, carrier molecules on the inner mitochondrial membrane pick up electrons from the hydrogen atoms carried by NAD and FAD. (Each hydrogen atom contains a hydrogen ion and an electron.) These carrier molecules transport the electrons through a series of enzyme-catalyzed oxidation-reduction reactions in the mitochondria.

Oxygen attraction

Oxygen plays a crucial role by attracting electrons along the chain of carriers in the transport system. During oxidation, a chemical compound loses electrons; during reduction, it gains electrons. These reactions release the energy contained in the electrons and generate ATP.

After passing through the electron transport system, the hydrogen ions produced in the Krebs cycle combine with oxygen to form water.

Liver and muscle cells

Both the liver and muscle cells help to regulate blood glucose levels.

> I convert glucose into glycogen or lipids.

Liver

When glucose levels exceed the body's immediate needs, hormones stimulate the liver to convert glucose into glycogen or lipids. Glycogen forms through glycogenesis; lipids form through lipogenesis.

Glucose shortage

When the blood glucose level drops excessively, the liver can form glucose by two processes:

☝ breakdown of glycogen to glucose through glycogenolysis

✌ synthesis of glucose from amino acids through gluconeogenesis.

Muscle cells

Muscle cells can convert glucose to glycogen for storage. However, they lack the enzymes to convert glycogen back to glucose when needed. During vigorous muscular activity, when oxygen requirements exceed the oxygen supply, muscle cells break down glycogen to yield lactic acid and energy. Lactic acid then builds up in the muscles, and muscle glycogen is depleted.

> During vigorous muscle activity, such as exercise, muscle cells break down glycogen, creating energy.

Glycogen returns

Some of the lactic acid diffuses from muscle cells, is transported to the liver, and is reconverted to glycogen. The liver converts the newly formed glycogen to glucose, which travels through the bloodstream to the muscles and reforms into glycogen.

Energize!

When muscle exertion stops, some of the accumulated lactic acid converts back to pyruvic acid. Pyruvic acid is oxidized completely to yield energy by means of the Krebs cycle and electron transport system.

Protein metabolism

Proteins are absorbed as amino acids and carried by the portal venous system to the liver and then throughout the body by blood. Absorbed amino acids mix with other amino acids in the body's amino acid pool. These other amino acids may be synthesized by the body from other substances such as *keto acids*, or they may be produced by protein breakdown.

> Some keto acids can enter the Krebs cycle directly by combining with an intermediate compound in the cycle.

Amino acid conversion

The body can't store amino acids. Instead, it converts them to protein or glucose or catabolizes them to provide energy. Before these changes can occur, however, amino acids must be transformed by deamination or transamination.

Deamination

In *deamination*, an amino group ($-NH_2$) splits off from an amino acid molecule to form a molecule of ammonia and one of keto acid. Most of the ammonia is converted to urea and excreted in the urine.

Transamination

In *transamination*, an amino group is exchanged for a keto group in a keto acid through the action of transaminase enzymes. During this process, the amino acid is converted to a keto acid and the original keto acid is converted to an amino acid.

Amino acid synthesis

Proteins are synthesized from 20 amino acids from the body's amino acid pool. (See *Essential and nonessential amino acids*, page 236.)

Essential and nonessential amino acids

Amino acids are the structural units of proteins. They're classified as essential or nonessential based on whether the human body can synthesize them. The 9 amino acids that can't be synthesized must be obtained from the diet. The other 11 can be synthesized and are therefore nonessential in the diet; however, they're needed for protein synthesis.

There are 20 amino acids: 9 essential and 11 nonessential.

Essential
- Histidine
- Isoleucine
- Leucine
- Lysine
- Methionine
- Phenylalanine
- Threonine
- Tryptophan
- Valine

Nonessential
- Alanine
- Arginine
- Asparagine
- Aspartic acid
- Cystine
- Glutamic acid
- Glycine
- Hydroxyproline
- Proline
- Serine
- Tyrosine

Power station

Amino acids not used for protein synthesis can be converted to keto acids and metabolized by the Krebs cycle and the electron transport system to produce energy.

Fat chance

Amino acids can also be converted to other nutrients such as fats. Those amino acids not used for protein synthesis may be converted to pyruvic acid and then to acetyl CoA. The acetyl CoA fragments condense to form long-chain fatty acids — a process that is the reverse of fatty acid breakdown. These fatty acids then combine with glycerol to form fats.

Going glucose

Amino acids can also be converted to glucose. They're first converted to pyruvic acid, which may then be converted to glucose.

Lipid metabolism

Until required for use as fuel, lipids are stored in adipose tissue within cells. When needed for energy, each fat molecule is hydrolyzed to glycerol and three molecules of fatty acids. Glycerol can be converted to pyruvic acid and then to acetyl CoA, which enters the Krebs cycle.

Ketone body formation

The liver normally forms ketone bodies from acetyl CoA fragments, derived largely from fatty acid catabolism. Acetyl CoA molecules yield three types of ketone bodies:

 acetoacetic acid

 beta-hydroxybutyric acid

 acetone.

Acetoacetic acid

Acetoacetic acid results from the combination of two acetyl CoA molecules and subsequent release of CoA from these molecules.

Beta-hydroxybutyric acid

Beta-hydroxybutyric acid forms when hydrogen is added to the oxygen atom in the acetoacetic acid molecule. The term beta indicates the location of the carbon atom containing the OH group.

Acetone

Acetone forms when the COOH group of acetoacetic acid releases CO_2. Muscle tissue, brain tissue, and other tissues oxidize these ketone bodies for energy.

Excessive ketone formation

Under certain conditions, the body produces more ketone bodies than it can oxidize for energy. Such conditions include fasting, starvation, and uncontrolled diabetes (in which the body can't break down glucose). The body must then use fat instead of glucose as its primary energy source.

Along with other tissues, I oxidize ketone bodies for energy.

Ketone cops

Use of fat instead of glucose for energy leads to an excess of ketone bodies. This condition disturbs normal acid-base balance and homeostatic mechanisms, leading to ketosis.

Lipid formation

Excess amino acids can be converted to fat through keto acid–acetyl CoA conversion. Glucose may be converted to pyruvic acid and then to acetyl CoA, which is converted into fatty acids and then fat (in much the same way that amino acids are converted into fat).

Hormonal regulation of metabolism

Normal body functions necessitate that blood glucose levels stay within a certain range.

Glucose, going up...

Hormones regulate the blood glucose level by stimulating the metabolic processes that restore a normal level in re-

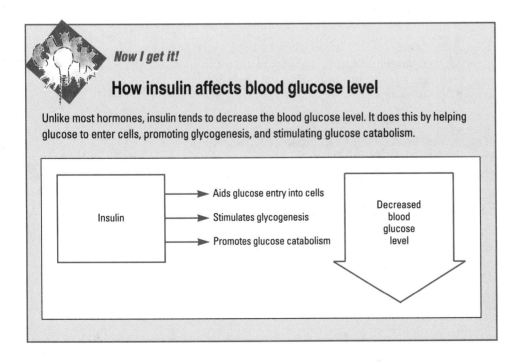

Now I get it!

How insulin affects blood glucose level

Unlike most hormones, insulin tends to decrease the blood glucose level. It does this by helping glucose to enter cells, promoting glycogenesis, and stimulating glucose catabolism.

Insulin
→ Aids glucose entry into cells
→ Stimulates glycogenesis
→ Promotes glucose catabolism

Decreased blood glucose level

sponse to blood glucose changes. (See *How insulin affects blood glucose level.*)

...and going down!

Insulin, produced by the pancreatic islet cells, is the only hormone that significantly reduces the blood glucose level. In addition to promoting cell uptake and use of glucose as an energy source, insulin promotes glucose storage as glycogen (glycogenesis) and lipids (lipogenesis).

Quick quiz

1. Which type of nutrient yields 9 kcal/g when used for energy?
    A.   Proteins
    B.   Carbohydrates
    C.   Lipids

Answer: C. Lipids are a concentrated form of fuel and yield 9 kcal/g.

2. Which hormone decreases the blood glucose level?
    A.   Epinephrine
    B.   Cortisol
    C.   Insulin

Answer: C. Insulin is the only hormone that significantly reduces blood glucose by aiding glucose entry into cells, stimulating glycogenesis, and promoting glucose catabolism.

3. Essential amino acids are:
    A.   the structural unit of protein that must be obtained from the diet.
    B.   the structural unit of protein that doesn't need to be obtained from the diet.
    C.   organic compounds that don't dissolve in water but do dissolve in alcohol and other organic solvents.

Answer: A. Essential amino acids can't be synthesized in the body and, therefore, must be obtained from the diet.

4.   Which vitamin is involved in prothrombin synthesis and other blood-clotting factors?
    A.   Vitamin K
    B.   Vitamin E
    C.   Vitamin B_{12}

Answer: A. Vitamin K is involved in liver synthesis of prothrombin and other blood-clotting factors.

Scoring

☆☆☆   If you answered all four questions correctly, hooray! You've digested a healthy dose of dense clinical material.

☆☆   If you answered two or three questions correctly, remarkable! You'll soon be a master of metabolism.

☆   If you answered fewer than two questions correctly, get energized! You've got four more quick quizzes to go!

Urinary system

Just the facts

In this chapter, you'll learn:

♦ major structures of the urinary system

♦ the functions of the kidneys, ureters, bladder, and urethra

♦ the role of hormone production in the kidneys.

Structures of the urinary system

The urinary system consists of two *kidneys*, two *ureters*, the *bladder*, and the *urethra*. Working together, these structures remove wastes from the body, help govern acid-base balance by retaining or excreting hydrogen ions, and regulate fluid and electrolyte balance.

Think of us as the body's plumbers. Together with the ureters, bladder, and urethra, we remove wastes from the body,...

...adjust acid-base balance, and regulate the balance of fluids and electrolytes.

Kidneys

The kidneys are bean-shaped, highly vascular organs. Each kidney consists of three regions:
- the renal cortex (outer region)
- the renal medulla (middle region)
- the renal pelvis (inner region). (See *Three regions: The kidney* and *A close look at the kidney.*)

The filtering station

The *renal cortex*, the outer region, contains blood-filtering mechanisms and is protected by a fibrous capsule and layers of fat.

Postcards of the pyramids

The *renal medulla*, the middle region of the kidney, contains 8 to 12 renal pyramids — striated wedges that are composed mostly of tubular structures. The tapered portion of each pyramid empties into a cuplike calyx. These calyces channel formed urine from the pyramids into the renal pelvis.

Keeping kidneys safe

The kidneys are protected in front by the contents of the abdomen and behind by the muscles attached to the vertebral column. A layer of fat surrounding each kidney offers further protection.

Perched on top

On top of each kidney lies an adrenal gland. These glands affect the renal system by influencing blood pressure as well as sodium and water retention in the kidneys.

No shortage of blood

Each kidney is supplied with blood by a renal artery, which subdivides into several branches when it enters the kidney. The kidneys are highly *vascular* (relating to or containing blood vessels), receiving about 20% of the blood pumped by the heart each minute.

All in a day's work

Together, these tissues allow the kidneys to perform their many functions. These functions include:
- elimination of wastes and excess ions (as urine)

Three regions: The kidney

The kidney has three regions:
- the *renal cortex* (outer region), which contains about 1.25 million renal tubules
- the *renal medulla* (middle region), which functions as the collecting chamber of the kidney
- the *renal pelvis* (inner region), which receives urine through the major calyces.

I'm a highly vascular organ, by the way. I receive about 20% of the blood pumped by the heart each minute.

Zoom in

A close look at the kidney

The kidneys are located in the lumbar area, with the right kidney situated slightly lower than the left to make room for the liver, which is just above it. The position of the kidneys shifts somewhat with changes in body position. Covering the kidneys are the true or fibrous capsule, perirenal fat, and renal fasciae.

Blood's cleansing journey

The kidneys receive waste-filled blood from the renal artery, which branches off the aorta. After passing through a complicated network of smaller blood vessels and nephrons, the filtered blood returns to the circulation by way of the renal vein, which empties into the inferior vena cava.

Continuing the cleanup

The kidneys excrete waste products that the nephrons remove from the blood; these excretions combine with other waste fluids (such as urea, creatine, phosphates, and sulfates) to form urine. An action called *peristalsis* (the circular contraction and relaxation of a tube-shaped structure) passes the urine through the ureters and into the urinary bladder. When the bladder has filled, nerves in the bladder wall relax the sphincter. In conjunction with a voluntary stimulus, this relaxation causes urine to pass into the urethra for elimination from the body.

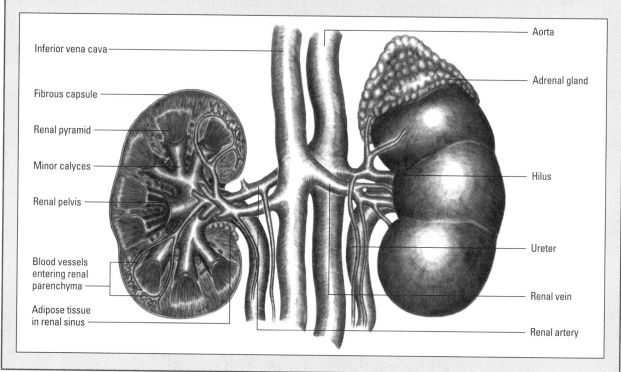

Inferior vena cava

Fibrous capsule

Renal pyramid

Minor calyces

Renal pelvis

Blood vessels entering renal parenchyma

Adipose tissue in renal sinus

Aorta

Adrenal gland

Hilus

Ureter

Renal vein

Renal artery

- blood filtration (by regulating chemical composition and volume of blood)
- maintenance of fluid-electrolyte and acid-base balances

• production of erythropoietin (a hormone that stimulates red blood cell production) and enzymes (such as renin, which governs blood pressure and kidney function)

• conversion of vitamin D to a more active form.

The nephron

Within the kidney, the *nephron* serves as the basic structural and functional unit. (See *Structure of the nephron.*)

Filter, reabsorb, and secrete

The nephrons perform two main functions:

• mechanically filtrating fluids, wastes, electrolytes, acids, and bases into the tubular system

• selectively reabsorbing and secreting ions.

Part and parcel

Each nephron consists of a tubular apparatus called the *glomerulus* as well as a collecting duct. The glomerulus is located inside a glomerular capsule, or *Bowman's capsule*, and consists of a cluster of capillaries.

Totally tubular

The nephron is divided into three portions. The portion nearest the glomerular capsule is the *proximal convoluted tubule*. The second portion, the *loop of Henle*, has an ascending and a descending limb. The third portion, the one farthest from the glomerular capsule, is the *distal convoluted tubule*. Its distal end joins the distal ends of neighboring nephrons, forming a larger collecting tubule.

Positively loopy

The glomeruli and proximal and distal tubules of the nephron are located in the renal cortex. The long loops of Henle, together with their accompanying blood vessels and collecting tubules, form the renal pyramids in the medulla.

Water, water everywhere

The proximal convoluted tubules have freely permeable cell membranes. This allows reabsorption of nearly all the filtrate's glucose, amino acids, metabolites, and electrolytes into nearby capillaries as well as allowing the circulation of large amounts of water.

Zoom in

Structure of the nephron

The *nephron* is the kidney's basic functional unit and the site of urine formation. The renal artery, a large branch of the abdominal aorta, carries blood to each kidney. Blood flows through the interlobular artery (running between the lobes of the kidneys) to the afferent arteriole, which conveys blood to the glomerulus, through which it passes into the efferent arteriole into the peritubular capillaries, venules, and the interlobular vein. The peritubular capillary network of vessels then supplies blood to the tubules of the nephron.

Time to concentrate

By the time the filtrate enters the descending limb of the loop of Henle, its water content has been reduced by 70%. At this point, the filtrate contains a high concentration of salts, chiefly sodium. As the filtrate moves deeper into the medulla and the loop of Henle, osmosis draws even more water into the extracellular spaces, further concentrating the filtrate.

Readjust and exit

After the filtrate enters the ascending limb, its concentration is readjusted by the transport of ions into the tubule. This transport continues until the filtrate enters the distal convoluted tubule.

Ureters

The *ureters* are fibromuscular tubes that connect each kidney to the bladder. Because the left kidney is higher than the right, the left ureter is usually slightly longer than the right ureter.

Triple protection

Each ureter is surrounded by a three-layered wall: the *mucosa*, the *muscularis*, and the *fibrous coat,* or outer layer, which holds the ureter in place. (See *Three layers: The ureter.*)

Riding the waves

The ureters act as conduits that carry urine from the kidneys to the bladder. Peristaltic waves occurring between one to five times each minute channel urine along the ureters toward the bladder.

Bladder

The *bladder* is a hollow, sphere-shaped, muscular organ in the pelvis. It lies anterior and inferior to the pelvic cavity and posterior to the *symphysis pubis* (the joint between the two pubic bones). Its function is to store urine. In a normal adult, bladder capacity ranges from 500 to 600 ml. If the amount of stored urine exceeds bladder capacity, the bladder distends above the symphysis pubis.

Three openings

The base of the bladder contains three openings that form a triangular area called the *trigone.* Two of the openings connect the bladder to the ureters, while the third connects the bladder to the urethra.

Three layers: The ureter

Each ureter has a three-layered wall:
• The *mucosa,* the innermost layer, contains the transitional epithelium.
• The *muscularis,* the middle layer, contains smooth muscle layers.
• Extensions of the *fibrous coat,* the outer layer, hold the ureter in place.

The ureters are the tubes that connect the kidneys to the bladder.

A matter of reflex

Urination results from *involuntary* (reflex) and voluntary (learned or intentional) processes.

When urine fills the bladder, parasympathetic nerve fibers in the bladder wall cause the bladder to contract and the *internal sphincter* (located at the internal urethral orifice) to relax. Then the cerebrum, in a voluntary reaction, causes the external sphincter to relax and urination to begin. This is called the *micturition* reflex.

> Urination is the result of two processes — one voluntary and one involuntary.

Urethra

The *urethra* is a small duct that channels urine from the bladder out of the body.

Straight to the outside

In the female, the urethra is embedded in the anterior wall of the vagina behind the symphysis pubis. The urethra connects the bladder with an external opening, or *urethral meatus*, located anterior to the vaginal opening.

Layers upon layers

The female urethra is composed of an inner layer of mucous membrane, a middle layer of spongy tissue, and an outer layer of muscle.

Passing through the prostate

In the male, the urethra passes vertically through the *prostate* gland and then extends through the urogenital diaphragm and the penis.

An additional function

The male urethra serves as a passageway for semen as well as urine.

Urine formation

Urine formation is one of the main functions of the urinary system. Urine formation results from three processes that occur in the nephrons: *glomerular filtration*, *tubular reabsorption*, and *tubular secretion*. (See *How the kidneys form urine*, page 248.)

Now I get it!

How the kidneys form urine

Urine formation consists of three steps: glomerular filtration, tubular reabsorption, and tubular secretion.

Step 1: Filter

As blood flows into the glomerulus, filtration occurs. In glomerular filtration, active transport from the proximal convoluted tubules leads to reabsorption of sodium (Na^+) and glucose into nearby circulation. *Osmosis* then causes water (H_2O) reabsorption.

Step 2: Reabsorb

In tubular reabsorption, a substance moves from the filtrate back from the distal convoluted tubules, into the peritubular capillaries. Active transport results in Na^+ reabsorption. The presence of antidiuretic hormone causes H_2O reabsorption.

Step 3: Secrete

In tubular secretion, a substance moves from the peritubular capillaries into the tubular filtrate. Peritubular capillaries then secrete ammonia (NH_3) and hydrogen (H^+) into the distal tubules by active transport.

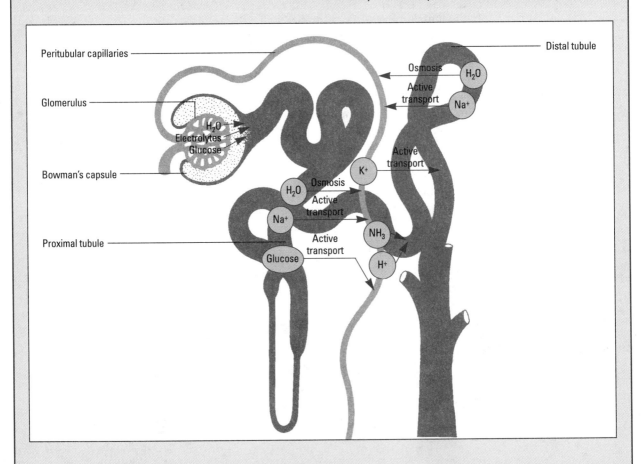

A mine of minerals

When formed, normal urine consists of sodium, chloride, potassium, calcium, magnesium, sulfates, phosphates, bicarbonates, uric acid, ammonium ions, creatinine, and *urobilinogen* (a derivative of bilirubin resulting from the action of intestinal bacteria).

A few leukocytes and red blood cells (and, in males, some spermatozoa) may enter the urine as it passes from the kidney to the ureteral orifice. If a person is taking drugs normally excreted in urine, the urine will contain those substances as well.

Kidneys in charge

The kidneys can vary the amount of substances reabsorbed and secreted in the nephrons, changing the composition of excreted urine.

Controlling the flow

Total daily urine output averages 720 to 2,400 ml, varying with fluid intake and climate. For example, after drinking a large volume of fluid, urine output increases as the body rapidly excretes excess water. If a person restricts or decreases water intake or has an excessive intake of sodium, urine output decreases as the body retains water to restore normal fluid concentration.

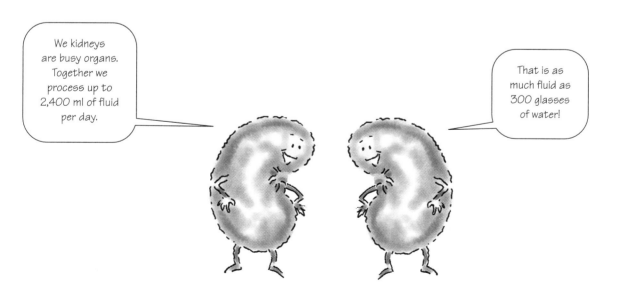

Hormones and the urinary system

Hormones play a large role in the urinary system, including helping the body to manage tubular reabsorption and secretion. Hormones affecting the urinary system include:

- antidiuretic hormone
- angiotensin I
- angiotensin II
- aldosterone
- erythropoietin.

Antidiuretic hormone

Antidiuretic hormone (ADH) regulates the levels of urine output. High levels of ADH increase water absorption and urine concentration, whereas lower levels of

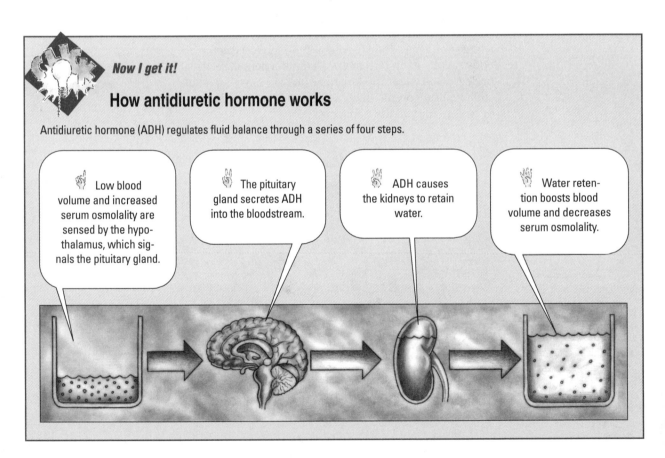

Now I get it!

How antidiuretic hormone works

Antidiuretic hormone (ADH) regulates fluid balance through a series of four steps.

Low blood volume and increased serum osmolality are sensed by the hypothalamus, which signals the pituitary gland.

The pituitary gland secretes ADH into the bloodstream.

ADH causes the kidneys to retain water.

Water retention boosts blood volume and decreases serum osmolality.

Now I get it!

Aldosterone production

Aldosterone (a hormone that helps to regulate fluid balance) is released by the adrenal gland through the actions of the renin-angiotensin system.

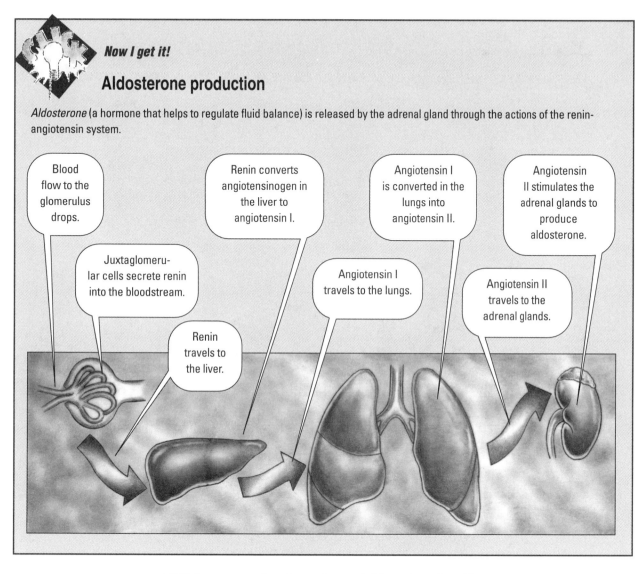

Blood flow to the glomerulus drops.

Juxtaglomeru-lar cells secrete renin into the bloodstream.

Renin travels to the liver.

Renin converts angiotensinogen in the liver to angiotensin I.

Angiotensin I travels to the lungs.

Angiotensin I is converted in the lungs into angiotensin II.

Angiotensin II travels to the adrenal glands.

Angiotensin II stimulates the adrenal glands to produce aldosterone.

ADH decrease water absorption and dilute the urine. (See *How antidiuretic hormone works.*)

Renin-angiotensin system

Renin is an enzyme secreted by the kidneys that circulates in the blood. It has no effect on blood pressure itself, but it leads to the formation of the hormone called *angiotensin I*. As it circulates through the lung, angiotensin I is converted into *angiotensin II* by *angiotensin-converting enzyme.* Angiotensin II exerts a powerful

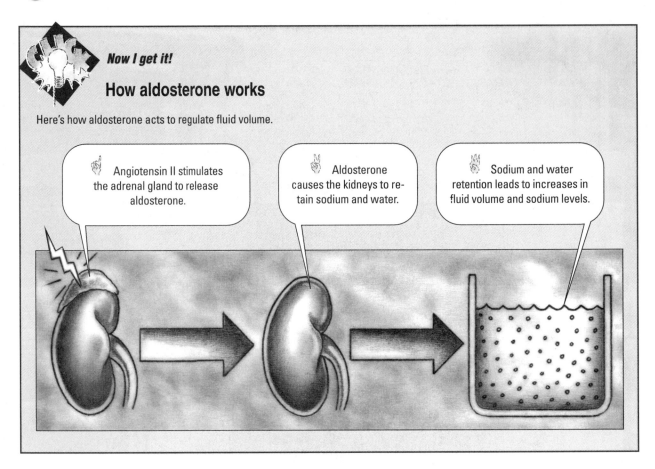

Now I get it!

How aldosterone works

Here's how aldosterone acts to regulate fluid volume.

> Angiotensin II stimulates the adrenal gland to release aldosterone.

> Aldosterone causes the kidneys to retain sodium and water.

> Sodium and water retention leads to increases in fluid volume and sodium levels.

constrictor effect on the arterioles and thus can raise blood pressure.

The best defense

A primary function of the renin-angiotensin system is to serve as a defense mechanism, maintaining the blood pressure in situations such as hemorrhage and extreme salt depletion. A low blood pressure and a low amount of salt passing through the kidneys are two of the three factors that stimulate the kidneys to release renin. (The third is the sympathetic nervous system.)

Aldosterone

The renin-angiotensin system has a second effect that makes it even more potent: It acts on the adrenal gland to release *aldosterone*. (See *Aldosterone production*, page 251.)

Blood pressure helper

Aldosterone is produced by the adrenal cortex. It facilitates tubular reabsorption by regulating sodium retention and helping to control potassium secretion by epithelial cells in the tubules.

When serum potassium levels rise, the adrenal cortex responds by increasing aldosterone secretion, which causes sodium retention, thereby raising the blood pressure. (See *How aldosterone works.*)

Other hormones

The kidneys secrete the hormone *erythropoietin* in response to low arterial oxygen tension. The hormone travels to the bone marrow, where it stimulates increased red blood cell production.

A balancing act

The kidneys also regulate calcium and phosphorus balance by filtering and reabsorbing approximately half of unbound serum calcium. In addition, the kidneys activate vitamin D_3, a compound that promotes intestinal calcium absorption and regulates phosphate excretion.

Quick quiz

1. In a normal adult, bladder capacity ranges from:
   A. 50 to 100 ml.
   B. 200 to 300 ml.
   C. 500 to 600 ml.

Answer: C. In a normal adult, bladder capacity ranges from 500 to 600 ml.

2. The left ureter is slightly longer than the right because the:
   A. left kidney is higher than the right.
   B. right kidney is higher than the left.
   C. left kidney performs more functions.

Answer: A. The left kidney is slightly higher than the right kidney. Therefore, the left ureter needs to be longer to reach the bladder.

3. Urination results from an involuntary and voluntary process. This process is called the:
- A. kidney process.
- B. micturition reflex.
- C. prostate reflex.

Answer: B. The micturition reflex is the signal system that occurs when urine fills the bladder. Parasympathetic nerve fibers in the bladder wall cause the bladder to contract and the internal sphincter to relax, which is followed by a voluntary relaxation of the external sphincter.

4. A person on a new health regimen has begun to drink at least 8 oz (236 ml) of water six to eight times per day. The kidneys react to this change by:
- A. producing aldosterone.
- B. secreting renin.
- C. increasing urine output.

Answer: C. Urine output typically varies with fluid intake and climate. After ingestion of a large volume of fluid, urine output increases as the body rapidly excretes excess water.

Scoring

☆☆☆   If you answered all four questions correctly, congratulations! You've got the kidneys (and ureters, bladder, and urethra) down cold.

☆☆   If you answered two or three questions correctly, not bad! You're learning to navigate the urinary system.

☆   If you answered fewer than two questions correctly, be of good cheer. There's still ample time for you to become an expert on anatomy and physiology!

Fluids, electrolytes, acids, and bases

Just the facts

In this chapter, you'll learn:

♦ how fluids are distributed throughout the body

♦ the role the kidneys play in electrolyte balance

♦ how the body compensates for acid-base imbalances

♦ how to identify major acid-base imbalances.

Fluid balance

The health and *homeostasis* (equilibrium of the various body functions) of human body depend on *fluid*, *electrolyte*, and *acid-base balance*. Factors that disrupt this balance, such as surgery, illness, and injury, can lead to potentially fatal changes in metabolic activity. (See *How the body gains and loses fluid*, page 256.)

> Fluids, electrolytes, acids, and bases — they've all got to be in balance for the body to function correctly.

Body shop

How the body gains and loses fluid

Each day, the body takes in fluid from the GI tract (in foods, liquids, and water of oxidation) and loses fluids through the skin, lungs, intestines (feces), and urinary tract (urine). This illustration shows the primary sites involved in fluid gains and losses as well as the amount of normal daily fluid intake and output.

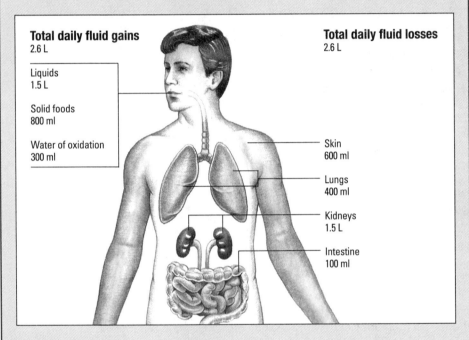

Total daily fluid gains
2.6 L

Liquids
1.5 L

Solid foods
800 ml

Water of oxidation
300 ml

Total daily fluid losses
2.6 L

Skin
600 ml

Lungs
400 ml

Kidneys
1.5 L

Intestine
100 ml

The four fluids

Body fluid is made up of water containing *solutes*, or dissolved substances, that are necessary for physiologic functioning. Solutes include electrolytes, glucose, amino acids, and other nutrients. There are four types of body fluids:
- *Intracellular fluid (ICF)* is found within the individual cells of the body.
- *Intravascular fluid (IVF)* is found within the plasma and lymphatic system.
- *Interstitial fluid (ISF)* is found in the loose tissue around cells.

Water weight

Water in the body exists in two major compartments that are separated by capillary walls and cell membranes. About two-thirds of the body's water is found within cells as intracellular fluid (ICF). The other third remains outside cells as extracellular fluid (ECF).

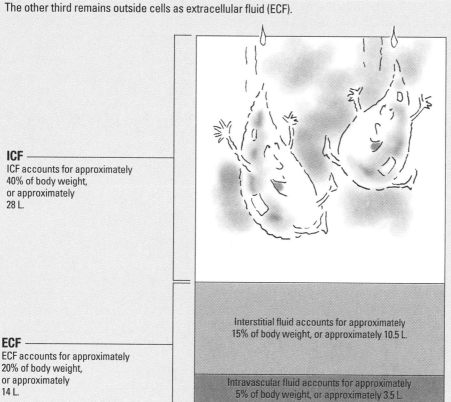

ICF
ICF accounts for approximately 40% of body weight, or approximately 28 L.

ECF
ECF accounts for approximately 20% of body weight, or approximately 14 L.

Interstitial fluid accounts for approximately 15% of body weight, or approximately 10.5 L.

Intravascular fluid accounts for approximately 5% of body weight, or approximately 3.5 L.

- *Extracellular fluid (ECF)*, found in the space between cells, is a combination of IVF and ISF. (See *Four types of body fluid.*)

Playing the percentages

ICF and ECF represent about 40% and 20%, respectively, of an adult's total body weight. (See *Water weight.*)

Fluids and their movement

Fluids in the body generally aren't found in pure forms. They're most often found in three different types of solutions: isotonic, hypotonic, and hypertonic.

Four types of body fluid

- Intracellular fluid (ICF) is found within the individual cells of the body.
- Intravascular fluid (IVF) is found within the plasma and lymphatic system.
- Interstitial fluid (ISF) is found in the loose tissue around cells.
- Extracellular fluid (ECF) a combination of IVF and ISF.

Isotonic stays put

An isotonic solution has the same solute concentration as another solution. For instance, if two fluids in adjacent compartments are equally concentrated, they're already in balance so the fluid inside each compartment stays put. No imbalance means no net fluid shift. For example, normal saline solution is considered isotonic because the concentration of sodium in the solution nearly equals the concentration of sodium in the blood.

Go low for hypo

A hypotonic solution has a lower solute concentration than another solution. For instance, one solution contains only a little sodium and another solution contains more. The first solution is hypotonic, compared with the second solution. As a result, fluid from the first solution — the hypotonic solution — would shift into the second solution until the two solutions had equal concentrations. (Remember, the body constantly strives to maintain a state of balance, or equilibrium.)

High for hyper

A hypertonic solution has a higher solute concentration than another solution. For instance, one solution contains a large amount of sodium and a second solution contains hardly any. The first solution is hypertonic compared to the second solution.

As a result, fluid would be drawn from the second solution into the first solution — the hypertonic solution — until the two solutions had equal concentrations. Again, the body strives to maintain a state of equilibrium.

Fluid movement within the cells

Just as the heart beats constantly, fluids and solutes move constantly within the body. That movement allows the body to maintain *homeostasis*, the constant state of balance the body seeks.

Solutes within the various compartments of the body (intracellular, interstitial, and intravascular) move through the membranes separating those compartments. The membranes are semipermeable, meaning that they allow some solutes to pass through but not others. Solutes move through membranes at the cellular level by diffusion (movement from an area of high concentration

Memory jogger

To help remember which fluid belongs to which compartment, keep in mind that *inter* means *between* (as in interval — between two events) and *intra* means within or inside (as in intravenous — inside a vein).

to an area of lower concentration), active transport, or osmosis.

Actively transporting

In *active transport*, solutes move from an area of lower concentration to an area of higher concentration. Think of active transport as swimming upstream. When a fish swims upstream, it has to expend energy.

ATP: Energy for moving

The energy required for a solute to move against concentration gradient comes from a substance called *adenosine triphosphate (ATP)*. Stored in all cells, ATP supplies energy for solute movement in and out of cells.

Some solutes, such as sodium and potassium, use ATP to move in and out of cells in a form of active transport called the *sodium-potassium pump*. Other solutes that require active transport to cross cell membranes include calcium ions, hydrogen ions, amino acids, and certain sugars.

Osmosis lets fluids through

Osmosis refers to the passive movement of fluid across a membrane from an area of lower solute concentration and comparatively more fluid into an area of higher solute concentration and comparatively less fluid. Osmosis stops when enough fluid has moved through the membrane to equalize the solute concentration on both sides of the membrane.

In with the good

Water normally enters the body from the GI tract. The body obtains about 1.5 L (1.6 qt) of water daily from consumed liquids and obtains another approximately 800 ml (26.6 oz) from solid foods, which may contain up to 97% water.

Oxidation of food in the body yields carbon dioxide (CO_2) and about 300 ml (10 oz) of water (water of oxidation).

Out with the bad

Water leaves the body through the skin (in perspiration), lungs (in expired air), GI tract (in feces), and urinary tract (in urine).

The major pipeline

The main route of water loss is urine excretion, which typically varies from 1 to 2.6 L daily.

Water losses through the skin and lungs amount to 1 L daily but may increase markedly with strenuous exertion, which predisposes a person to dehydration.

Don't interrupt

In a healthy body, fluid gains match fluid losses to maintain proper physiologic functioning. However, interruption or dysfunction of one or both of the mechanisms that regulate fluid balance — thirst and the countercurrent mechanism — can lead to a fluid imbalance.

I'm parched!

Thirst — the conscious desire for water — is the primary regulator of fluid intake. When the body becomes dehydrated, ECF volume is reduced, causing an increase in sodium concentration and osmolarity.

When the sodium concentration reaches about 2 mEq/L above normal, neurons of the thirst center in the hypothalamus are stimulated. The brain then directs motor neurons to satisfy thirst, causing the person to drink enough fluid to restore ECF to normal.

> The body receives its daily fluids from consumed liquids as well as from solid foods, which can be up to 97% water.

What comes in must go out

Through the countercurrent mechanism, the kidneys regulate fluid output by modifying urine concentration — that is, by excreting urine of greater or lesser concentration, depending on fluid balance.

Electrolyte balance

Electrolytes are substances that *dissociate* (break up) into electrically charged particles, called *ions*, when dissolved in water. Adequate amounts of each major electrolyte and a proper balance among the electrolytes must be present to maintain normal physiologic functioning.

All charged up

Ions may be positively charged *cations* or negatively charged *anions*. Major cations include sodium, potassium, calcium, and magnesium. Major anions include chloride, bicarbonate (HCO_3^-), and phosphate.

Normally, the electrical charges of cations balance the electrical charges of anions, keeping body fluids electrically neutral. Blood plasma contains slightly more electrolytes than does ISF.

Shh! We're concentrating

Because ions are present in such low concentrations in body fluids, they're usually expressed in milliequivalents per liter (mEq/L). ICF and ECF cells are permeable to different substances; therefore, these compartments normally have different electrolyte compositions. (See *Electrolyte composition in ICF and ECF.*)

A delicate balance

Electrolytes profoundly affect the body's water distribution, osmolarity, and acid-base balance. Numerous mechanisms within the body help maintain electrolyte balance. Dysfunction or interruption of any of these mechanisms can produce an electrolyte imbalance.

Electrolyte composition in ICF and ECF

This chart presents the electrolyte compositions of intracellular fluid (ICF) and extracellular fluid (ECF).

Electrolyte	ICF	ECF
Sodium	10 mEq/L	136 to 146 mEq/L
Potassium	140 mEq/L	3.6 to 5 mEq/L
Calcium	10 mEq/L	4.5 to 5.8 mEq/L
Magnesium	40 mEq/L	1.6 to 2.2 mEq/L
Chloride	4 mEq/L	96 to 106 mEq/L
Bicarbonate	10 mEq/L	24 to 28 mEq/L
Phosphate	100 mEq/L	1 to 1.5 mEq/L

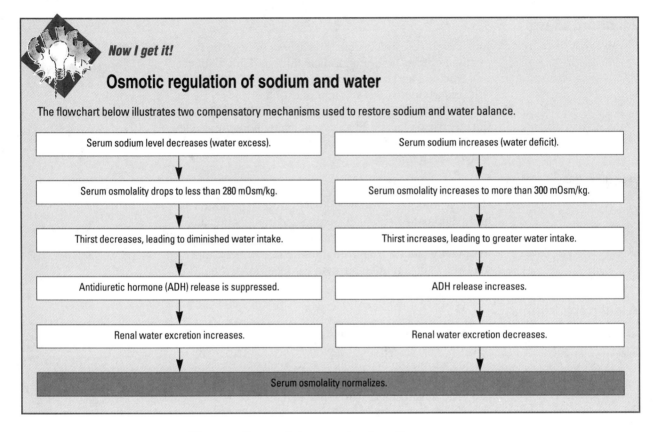

Now I get it!

Osmotic regulation of sodium and water

The flowchart below illustrates two compensatory mechanisms used to restore sodium and water balance.

Serum sodium level decreases (water excess).	Serum sodium increases (water deficit).
↓	↓
Serum osmolality drops to less than 280 mOsm/kg.	Serum osmolality increases to more than 300 mOsm/kg.
↓	↓
Thirst decreases, leading to diminished water intake.	Thirst increases, leading to greater water intake.
↓	↓
Antidiuretic hormone (ADH) release is suppressed.	ADH release increases.
↓	↓
Renal water excretion increases.	Renal water excretion decreases.

Serum osmolality normalizes.

Here are the regulatory mechanisms for common electrolytes:

• The kidneys and a hormone called aldosterone are the chief sodium regulators. The small intestine absorbs sodium readily from food, and the skin and kidneys excrete sodium. (See *Osmotic regulation of sodium and water.*)

• The kidneys also regulate potassium through aldosterone action. Most potassium is absorbed from food in the GI tract; normally, the amount excreted in urine equals dietary potassium intake.

• Calcium in the blood is typically in equilibrium with calcium salts in bone. Parathyroid hormone (PTH) is the main regulator of calcium, controlling both calcium uptake from the GI tract and calcium excretion by the kidneys.

• Magnesium is governed by aldosterone, which controls renal magnesium reabsorption. Absorbed from the GI tract, magnesium is excreted in urine, breast milk, and saliva.

• The kidneys also regulate chloride. Chloride ions move in conjunction with sodium ions.
• The kidneys regulate HCO_3^-, excreting, absorbing, or forming it. HCO_3^-, in turn, plays a vital part in acid-base balance.
• The kidneys regulate phosphate. Absorbed from food, phosphate is incorporated with calcium in bone. Along with calcium, PTH governs phosphate levels.

I'm a star of the electrolyte scene. I help regulate potassium, chloride, bicarbonate, and phosphate.

Acid-base balance

Physiologic survival requires *acid-base balance*, a stable concentration of hydrogen ions in body fluids.

An acid remark

An *acid* is a substance that yields hydrogen ions when *dissociated* (changed from a complex to a simpler compound) in solution. A strong acid dissociates almost completely, releasing a large number of hydrogen ions.

A base reply

A *base* dissociates in water, releasing ions that can combine with hydrogen ions. Like a strong acid, a strong base dissociates almost completely, releasing many ions.

The hydrogen ion concentration of a fluid determines whether it's *acidic* or *basic* (alkaline). A *neutral* solution, such as pure water, dissociates only slightly. (See *Understanding pH*, page 264.)

Making hydrogen ions

The body produces acids, thus yielding hydrogen ions, through the following mechanisms:
• Protein catabolism yields nonvolatile acids, such as sulfuric, phosphoric, and uric acids.
• Fat oxidation produces acid ketone bodies.
• Anaerobic glucose catabolism produces lactic acid.
• Intracellular metabolism yields CO_2 as a by-product; CO_2 dissolves in body fluids to form carbonic acid (H_2CO_3).

A balancing act

Normally, even with the production of these acids, the body's pH control mechanism is so effective that blood pH stays within a narrow

Will a fluid be acidic or basic? It all depends on the hydrogen ion concentration.

Now I get it!

Understanding pH

Hydrogen (H) ion concentration is commonly expressed as pH, which indicates the degree of acidity or alkalinity of a solution:
• A pH of 7 indicates neutrality or equal amounts of H and hydroxyl (OH^-) ions.
• An acidic solution contains more H ions than OH^- ions; its pH is less than 7.
• An alkaline solution contains fewer H ions than OH^- ions; its pH exceeds 7.
    Overall, as hydrogen concentration increases, pH goes down.

1 means 10
Because pH is an exponential expression, a change of one pH unit reflects a 10-fold difference in actual H ion concentration. For instance, a solution with a pH of 7 has 10 times more H ions than a solution with a pH of 8.

range: 7.35 to 7.45. This acid-base balance is maintained by buffer systems and the lungs and kidneys, which neutralize and eliminate acids as rapidly as they are formed.

Dysfunction or interruption of a buffer system or other governing mechanism can cause an acid-base imbalance. (See *Understanding respiratory and metabolic alkalosis and acidosis.*)

Buffer systems

Buffer systems reduce the effect of an abrupt change in hydrogen ion concentration by converting a strong acid or base (which normally would dissociate completely) into a weak acid or base (which releases fewer hydrogen ions).

Buffer systems that help maintain acid-base balance include:
• sodium bicarbonate–carbonic acid
• phosphate
• protein.

(Text continues on page 269.)

Now I get it!

Understanding respiratory and metabolic alkalosis and acidosis

What happens in respiratory alkalosis

Step 1

When pulmonary ventilation increases above the amount needed to maintain normal carbon dioxide (CO_2) levels, excessive amounts of CO_2 are exhaled. This causes hypocapnia (a fall in partial pressure of arterial carbon dioxide [$Paco_2$]), which leads to a reduction in carbonic acid (H_2CO_3) production, a loss of hydrogen (H) ions and bicarbonate (HCO_3^-) ions, and a subsequent rise in pH. Look for a pH level above 7.45, a $Paco_2$ level below 35 mm Hg, and an HCO_3^- level below 22 mEq/L. (as shown at right).

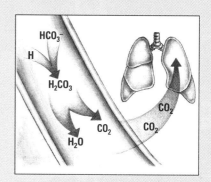

Step 2

In defense against the rising pH, H ions are pulled out of the cells and into the blood in exchange for potassium (K) ions. The H ions entering the blood combine with HCO_3^- ions to form H_2CO_3, which lowers the pH. Look for a further decrease in HCO_3^- levels, a fall in pH, and a fall in serum K levels (hypokalemia).

Step 3

Hypocapnia stimulates the carotid and aortic bodies and the medulla, which causes an increase in heart rate without an increase in blood pressure (as shown at right). Look for angina, electrocardiogram changes, restlessness, and anxiety.

Step 4

Simultaneously, hypocapnia produces cerebral vasoconstriction, which prompts a reduction in cerebral blood flow. Hypocapnia also overexcites the medulla, pons, and other parts of the autonomic nervous system. Look for increasing anxiety, diaphoresis, dyspnea, alternating periods of apnea and hyperventilation, dizziness, and tingling in the fingers or toes.

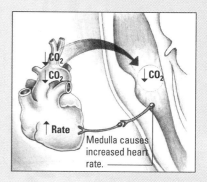

Step 5

When hypocapnia lasts more than 6 hours, the kidneys increase secretion of HCO_3^- and reduce excretion of H (as shown at right). Periods of apnea may result if the pH remains high and the $Paco_2$ remains low. Look for slowing of the respiratory rate, hypoventilation, and Cheyne-Stokes respirations.

Step 6

Continued low $Paco_2$ increases cerebral and peripheral hypoxia from vasoconstriction. Severe alkalosis inhibits calcium (Ca) ionization, which in turn causes increased nerve excitability and muscle contractions. Eventually, the alkalosis overwhelms the central nervous system and the heart. Look for decreasing level of consciousness, hyperreflexia, carpopedal spasm, tetany, arrhythmias, seizures, and coma.

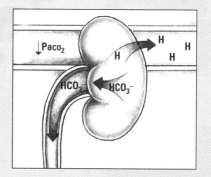

(continued)

Understanding respiratory and metabolic alkalosis and acidosis (continued)

What happens in respiratory acidosis

Step 1
When pulmonary ventilation decreases, retained CO_2 combines with water (H_2O) to form H_2CO_3 in larger-than-normal amounts. The H_2CO_3 dissociates to release free H ions and HCO_3^- ions. The excessive H_2CO_3 causes a drop in pH. Look for a $Paco_2$ level above 45 mm Hg and a pH level below 7.35.

Step 2
As the pH level falls, 2,3-diphosphoglycerate (2,3-DPG) increases in the red blood cells and causes a change in hemoglobin (Hb) that makes the Hb release oxygen (O_2). The altered Hb, now strongly alkaline, picks up H ions and CO_2 thus eliminating some of the free H ions and excess CO_2 (as shown at right). Look for decreased arterial oxygen saturation.

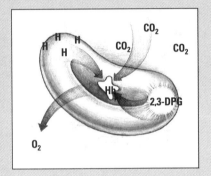

Step 3
Whenever $Paco_2$ increases, CO_2 builds up in all tissues and fluids, including cerebrospinal fluid and the respiratory center in the medulla. The CO_2 reacts with H_2O to form H_2CO_3, which then breaks into free H ions and HCO_3^- ions. The increased amount of CO_2 and free H ions stimulate the respiratory center to increase the respiratory rate. An increased respiratory rate expels more CO_2 and helps to reduce the CO_2 level in the blood and other tissues. Look for rapid, shallow respirations and a decreasing $Paco_2$.

Step 4
Eventually, CO_2 and H ions cause cerebral blood vessels to dilate, which increases blood flow to the brain. That increased flow can cause cerebral edema and depress central nervous system activity. Look for headache, confusion, lethargy, nausea, or vomiting. As respiratory mechanisms fail, the increasing $Paco_2$ stimulates the kidneys to retain HCO_3^- and sodium (Na) ions and to excrete H ions, some of which are excreted in the form of ammonium (NH_4). The additional HCO_3^- and Na combine to form extra sodium bicarbonate ($NaHCO_3^-$), which is then able to buffer more free H ions (as shown at right). Look for increased acid content in the urine, increasing serum pH and HCO_3^- levels, and shallow, depressed respirations.

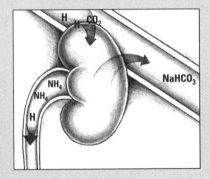

Step 5
As the concentration of H ions overwhelms the body's compensatory mechanisms, the H ions move into the cells and potassium (K) ions move out. A concurrent lack of O_2 causes an increase in the anaerobic production of lactic acid, which further skews the acid-base balance and critically depresses neurologic and cardiac functions. Look for hyperkalemia, arrhythmias, increased $Paco_2$, decreased partial pressure of arterial oxygen, decreased pH, and decreased level of consciousness.

Understanding respiratory and metabolic alkalosis and acidosis (continued)

What happens in metabolic alkalosis

Step 1
As HCO_3^- ions start to accumulate in the body, chemical buffers (in extracellular fluid and cells) bind with the ions. No signs are detectable at this stage.

Step 2
Excess HCO_3^- ions that don't bind with chemical buffers elevate serum pH levels, which in turn depress chemoreceptors in the medulla. Depression of those chemoreceptors causes a decrease in respiratory rate, which increases the $Paco_2$. The additional CO_2 combines with H_2O to form H_2CO_3 (as shown at right). Note: Lowered oxygen levels limit respiratory compensation. Look for a serum pH level above 7.45, an HCO_3^- level above 26 mEq/L, a rising $Paco_2$, and slow, shallow respirations.

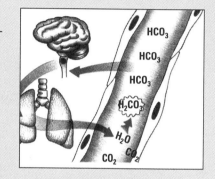

Step 3
When the HCO_3^- level exceeds 28 mEq/L, the renal glomeruli can no longer reabsorb excess HCO_3^-. That excess HCO_3^- is excreted in the urine; H ions are retained. Look for alkaline urine and pH and HCO_3^- levels that slowly return to normal.

Step 4
To maintain electrochemical balance, the kidneys excrete excess Na ions, H_2O, and HCO_3^- (as shown at right). Look for polyuria initially, then signs of hypovolemia, including thirst and dry mucous membranes.

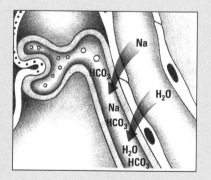

Step 5
Lowered H ion levels in the extracellular fluid cause the ions to diffuse out of the cells. To maintain the balance of charge across the cell membrane, extracellular K ions move into the cells. Look for signs of hypokalemia: anorexia, muscle weakness, loss of reflexes, and others.

Step 6
As H ion levels decline, Ca ionization decreases. That decrease in ionization makes nerve cells more permeable to Na ions. Na ions moving into nerve cells stimulate neural impulses and produce overexcitability of the peripheral and central nervous systems. Look for tetany, belligerence, irritability, disorientation, and seizures.

As metabolic alkalosis progresses, it ultimately leads to tetany, belligerence, irritability, disorientation, and seizures.

(continued)

Understanding respiratory and metabolic alkalosis and acidosis *(continued)*

What happens in metabolic acidosis

Step 1
As H ions start to accumulate in the body, chemical buffers (plasma HCO_3^- and proteins) in the cells and extracellular fluid bind with them (as shown at right). No signs are detectable at this stage.

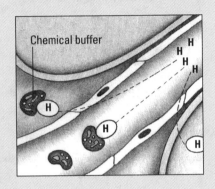

Chemical buffer

Step 2
Excess H ions that the buffers can't bind with decrease the pH and stimulate chemoreceptors in the medulla to increase the respiratory rate. The increased respiratory rate lowers the $Paco_2$, which allows more H ions to bind with HCO_3^- ions. Respiratory compensation occurs within minutes but isn't sufficient to correct the imbalance (see middle illustration). Look for a pH level below 7.35, an HCO_3^- level below 22 mEq/L, a decreasing $Paco_2$ level, and rapid, deeper respirations.

Step 3
Healthy kidneys try to compensate for the acidosis by secreting excess H ions into the renal tubules. Those ions are buffered by phosphate or ammonia and then are excreted into the urine in the form of a weak acid. Look for acidic urine.

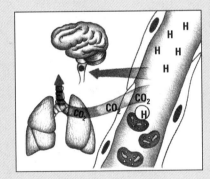

Step 4
Each time an H ion is secreted into the renal tubules, an Na ion and an HCO_3^- ion are absorbed from the tubules and returned to the blood. Look for pH and HCO_3^- levels that slowly return to normal.

Step 5
Excess H ions in the extracellular fluid diffuse into cells. To maintain the balance of the charge across the membrane, the cells release K ions into the blood (as shown at right). Look for signs of hyperkalemia, including colic and diarrhea, weakness or flaccid paralysis, tingling and numbness in the extremities, bradycardia, a tall T wave, a prolonged PR interval, and a wide QRS complex.

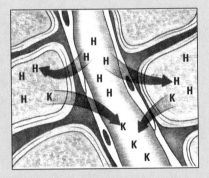

Step 6
Excess H ions alter the normal balance of K, Na, and Ca ions, leading to reduced excitability of nerve cells. Look for signs and symptoms of progressive central nervous system depression, including lethargy, dull headache, confusion, stupor, and coma.

One from the kidneys, one from the lungs

The *sodium bicarbonate–carbonic acid* buffer system is the major buffer in ECF. Sodium bicarbonate ($NaHCO_3^-$) concentration is regulated by the kidneys, and H_2CO_3 concentration is regulated by the lungs. Both components of this buffer are replenished continually. As a result of the buffering action, the strong base (sodium hydroxide [NaOH]) is replaced by $NaHCO_3^-$ and water (H_2O). NaOH dissociates almost completely and releases large amounts of hydroxyl (OH^-). If a strong acid is added, the opposite occurs.

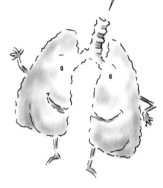

Along with the kidneys, I help blood pH stay in the normal range: 7.35 to 7.45...

Finesse with phosphate

A *phosphate* buffer system works by regulating pH of fluids as they pass through the kidneys. It's also important in ECF. Its acidic component is *sodium dihydrogen phosphate* (NaH_2PO_4); its alkaline component is *sodium monohydrogen phosphate* (Na_2HPO_4).

Absorbing ions

In the *protein* buffer system, intracellular proteins absorb hydrogen ions generated by the body's metabolic processes.

Lungs

The protein buffer system changes the pH of the blood in 3 minutes or less by changing the breathing rate. A decreased respiratory rate decreases the exchange and release of CO_2, there is less hydrogen (H^+), and the pH rises. Present in all acids, H ions are protons that can be added to or removed from a solution to change the pH.

Respiration plays a crucial role in controlling pH. The lungs excrete CO_2 and regulate the H_2CO_3 content of the blood. H_2CO_3 is derived from CO_2 and H_2O that are released as by-products of cellular metabolic activity.

...I do it by excreting CO_2 and regulating the blood's H_2CO_3 content.

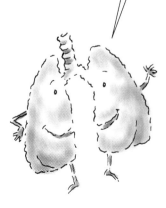

Stick to the formula

CO_2 is soluble in blood plasma. Some of the dissolved gas reacts with H_2O to form H_2CO_3, a weak acid that partially breaks apart to form H and HCO_3^- ions. These three substances are in equilibrium, as reflected in the following formula:

$$CO_2 + H_2O \leftrightarrow H_2CO_3 \leftrightarrow H^+ + HCO_3^-.$$

CO_2 dissolved in plasma is in equilibrium with CO_2 in the lung alveoli (expressed as a partial pressure [Pco_2]). Thus, an equilibrium exists between alveolar Pco_2 and the various forms of CO_2 present in the plasma, as expressed by the following formula:

$$Pco_2 \leftrightarrow CO_2 + H_2O \leftrightarrow H_2CO_3 \leftrightarrow H + HCO_3^-.$$

Don't hold your breath!

A change in the rate or depth of respirations can alter the CO_2 content of alveolar air and the alveolar Pco_2. A change in alveolar Pco_2 produces a corresponding change in the amount of H_2CO_3 formed by dissolved CO_2. In turn, these changes stimulate the respiratory center to modify respiratory rate and depth.

An increase in alveolar Pco_2 raises the blood concentration of CO_2 and H_2CO_3. This, in turn, stimulates the respiratory center to increase respiratory rate and depth. As a result, alveolar Pco_2 decreases, which leads to a corresponding drop in the H_2CO_3 and CO_2 concentrations in blood. (See *How respiratory mechanisms affect blood pH.*)

A decrease in respiratory rate and depth has the reverse effect; it raises alveolar Pco_2, which in turn triggers an increase in the blood's CO_2 and H_2CO_3 concentrations.

Kidneys

In addition to excreting various acid waste products, the kidneys help manage acid-base balance by regulating the blood's HCO_3^- concentration. They do so by permitting HCO_3^- reabsorption from tubular filtrate and by forming additional HCO_3^- to replace that used in buffering acids.

Renal tubular ion secretion

Recovery and formation of HCO_3^- in the kidneys depend on hydrogen ion secretion by the renal tubules in exchange for sodium (Na) ions. Na ions are then simultaneously reabsorbed into the circulation from the tubular filtrate.

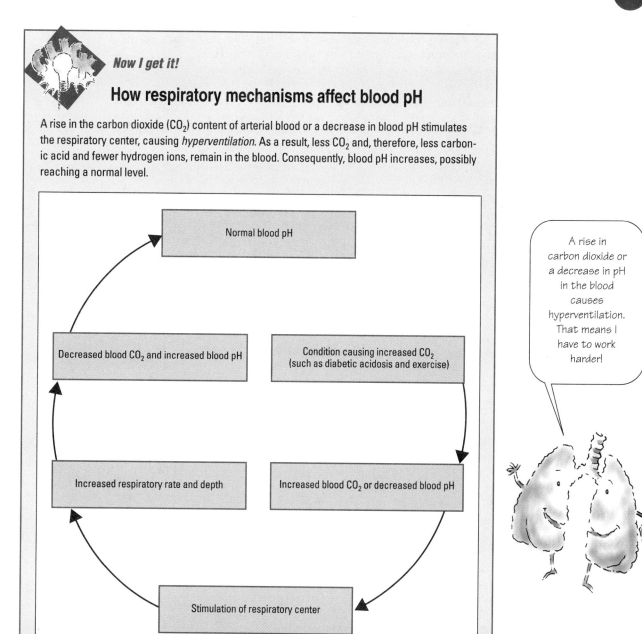

Now I get it!

How respiratory mechanisms affect blood pH

A rise in the carbon dioxide (CO_2) content of arterial blood or a decrease in blood pH stimulates the respiratory center, causing *hyperventilation*. As a result, less CO_2 and, therefore, less carbonic acid and fewer hydrogen ions, remain in the blood. Consequently, blood pH increases, possibly reaching a normal level.

Normal blood pH

Decreased blood CO_2 and increased blood pH

Condition causing increased CO_2 (such as diabetic acidosis and exercise)

Increased respiratory rate and depth

Increased blood CO_2 or decreased blood pH

Stimulation of respiratory center

A rise in carbon dioxide or a decrease in pH in the blood causes hyperventilation. That means I have to work harder!

Influential enzyme

The enzyme *carbonic anhydrase* influences tubular epithelial cells to form H_2CO_3 from CO_2 and H_2O. H_2CO_3

quickly dissociates into hydrogen and HCO_3^- ions. H ions enter the tubular filtrate in exchange for Na ions; HCO_3^- ions enter the bloodstream along with the Na ions that have been absorbed from the filtrate. HCO_3^- is then reabsorbed from the tubular filtrate.

Ions hanging out together

Each H ion secreted into the tubular filtrate joins with an HCO_3^- ion to form H_2CO_3, which rapidly dissociates into CO_2 and H_2O. The CO_2 diffuses into the tubular epithelial cell, where it can combine with more H_2O and lead to the formation of more H_2CO_3.

Bicarbonate reabsorption

The remaining H_2O molecule in the tubular filtrate is eliminated in the urine. As each H ion enters the tubular filtrate to combine with an HCO_3^- ion, an HCO_3^- ion in the tubular epithelial cell diffuses into the circulation. This process is termed HCO_3^- reabsorption. (However, the HCO_3^- ion that enters the circulation isn't the same one as in the tubular filtrate.)

Formation of ammonia and phosphate salts

To form more HCO_3^-, the kidneys must secrete additional H ions in exchange for sodium ions. For the renal tubules to continue secreting H ions, the excess ions must combine with other substances in the filtrate and be excreted. Excess H ions in the filtrate may combine with *ammonia* (NH_3) produced by the renal tubules) or with phosphate salts (present in the tubular filtrate).

More ions on the move

NH_3 forms in the tubular epithelial cells by removal of the amino groups from *glutamine* (an amino acid derivative) and from amino acids delivered to the tubular epithelial cells from the circulation. After diffusing into the filtrate, NH_3 joins with the secreted H ions, forming ammonium ions (NH_4^+). These ions are excreted in the urine with chloride and other anions; each secreted NH_3 molecule eliminates one H ion in the filtrate.

At the same time, Na ions that have been absorbed from the filtrate and exchanged for H ions enter the circulation, as does the HCO_3^- formed in the tubular epithelial cells. Some secreted H ions combine with Na_2HPO_4, a *disodium phosphate salt* in the tubular filtrate. Each of the secreted H ions that joins with the disodium salt changes it to the monosodium salt NaH_2PO_4. The Na ion released in this reaction is absorbed into the circulation along with a newly formed HCO_3^- ion.

Factors affecting bicarbonate formation

The rate of HCO_3^- formation by renal tubular epithelial cells is affected by two factors:
- the amount of dissolved CO_2 in the plasma
- the potassium content of the tubular cells.

 If the plasma CO_2 level rises, renal tubular cells form more HCO_3^-. Increased plasma CO_2 encourages greater H_2CO_3 formation by renal tubular cells.

Chain reaction

Partial dissociation of H_2CO_3 results in more H ions for excretion into the tubular filtrate and additional HCO_3^- ions for entry into the circulation. This, in turn, increases the plasma HCO_3^- level and reduces the plasma level of dissolved CO_2 toward normal. If the plasma CO_2 level decreases, renal tubular cells form less H_2CO_3.

 Because fewer H ions are formed and excreted, fewer HCO_3^- ions enter the circulation. The plasma HCO_3^- level then falls accordingly.

Sphere of influence

The potassium (K) content of renal tubular cells also helps regulate plasma HCO_3^- concentration by affecting the rate at which the renal tubules secrete H ions. Tubular cell K content and H ion secretion are interrelated; K and H ions are secreted at rates that vary inversely. H ion secretion increases if tubular secretion of K ions falls; H ion secretion declines if tubular secretion of K ions increases.

 For each H ion secreted into the tubular filtrate, an additional HCO_3^- ion enters the blood plasma. Consequently, increased tubular secretion of H ions leads to a rise in the plasma HCO_3^- content.

 Depletion of body K causes more HCO_3^- to enter the circulation; the plasma HCO_3^- level then rises above normal. When the body contains excess K, the tubules excrete more K. As a result, fewer H ions are secreted, less HCO_3^- forms, and the plasma HCO_3^- concentration decreases.

Potassium secretion by tubular epithelial cells decreases and hydrogen ion secretion rises in patients with potassium depletion from vomiting or diarrhea.

Quick quiz

1. A solution with a pH less than 7 is considered:
    A. acidic.
    B. alkaline.
    C. solute.

Answer: A. A solution with a pH less than 7 contains more H ions than OH$^-$ ions and is considered acidic.

2. The body compensates for chronic respiratory alkalosis by developing:
    A. metabolic alkalosis.
    B. respiratory acidosis.
    C. metabolic acidosis.

Answer: C. The body compensates for chronic respiratory alkalosis by developing metabolic acidosis.

3. The two factors that affect the rate of HCO$_3^-$ formation by renal tubular epithelial cells are:
    A. amount of aldosterone in the system; urine production.
    B. amount of dissolved CO$_2$ in the plasma; the potassium content of the tubular cells.
    C. amount of dissolved potassium in the plasma; the CO$_2$ content of the tubular cells.

Answer: B. The rate of HCO$_3^-$ formation by renal tubular epithelial cells is affected by the amount of dissolved CO$_2$ in the plasma and the potassium content of the tubular cells.

Scoring

☆☆☆ If you answered all three questions correctly, give yourself a pat on the back! You've got all your knowledge in the proper balance.

☆☆ If you answered two questions correctly, good for you! You're benefiting from all the right buffer systems.

☆ If you answered fewer than two questions correctly, check out the chapter again. It may take time to find your equilibrium.

Reproductive system

Just the facts

In this chapter, you'll learn:

♦ the anatomic structure and functions of the male and female reproductive systems

♦ male hormone production and how it affects sexual development

♦ female hormone production and how it affects menstruation

♦ the female breast and its function.

Male reproductive system

Anatomically, the main distinction between the male and female is the presence of conspicuous external genitalia in the male. In contrast, the major reproductive organs of the female lie within the pelvic cavity.

Here's the main difference — males have major external genitalia...

...but most female reproductive organs are inside the pelvic cavity.

Making introductions

The male reproductive system consists of the organs that produce, transfer, and introduce mature sperm into the female reproductive tract, where fertilization occurs. (See *Structures of the male reproductive system.*)

Extra work

In addition to supplying male sex cells (spermatogenesis), the male reproductive system plays a part in the secretion of male sex hormones.

Zoom in

Structures of the male reproductive system

The male reproductive system consists of the penis, the scrotum and its contents, the prostate gland, and the inguinal structures.

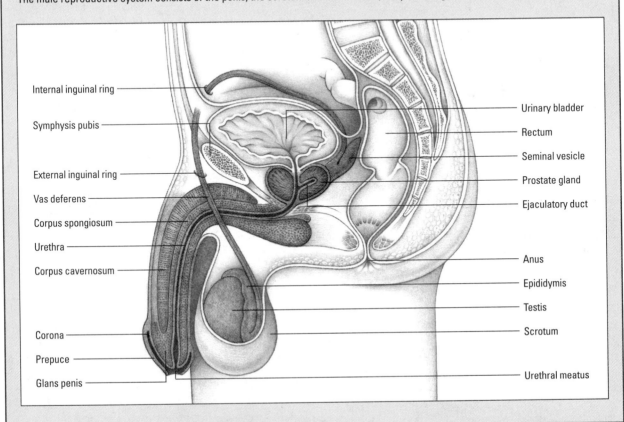

Internal inguinal ring

Symphysis pubis

External inguinal ring

Vas deferens

Corpus spongiosum

Urethra

Corpus cavernosum

Corona

Prepuce

Glans penis

Urinary bladder

Rectum

Seminal vesicle

Prostate gland

Ejaculatory duct

Anus

Epididymis

Testis

Scrotum

Urethral meatus

Penis

The organ of copulation, the *penis* deposits sperm in the female reproductive tract and acts as the terminal duct for the urinary tract. The penis also serves as the means for urine elimination. It consists of an attached root, a free shaft, and an enlarged tip.

What's inside

Internally, the cylinder-shaped penile shaft consists of three columns of *erectile tissue* bound together by *heavy fibrous tissue*. Two *corpora cavernosa* form the major part of the penis. On the underside, the *corpus spongiosum* encases the urethra. Its enlarged proximal end forms the bulb of the penis.

The *glans penis*, at the distal end of the shaft, is a cone-shaped structure formed from the corpus spongiosum. Its lateral margin forms a ridge of tissue known as the *corona*. The glans penis is highly sensitive to sexual stimulation.

What's outside

Thin, loose skin covers the penile shaft. The *urethral meatus* opens through the glans to allow urination and ejaculation.

In a different vein

The penis receives blood through the *internal pudendal artery*. Blood then flows into the corpora cavernosa through the penile artery. Venous blood returns through the *internal iliac vein* to the *vena cava*.

Scrotum

The penis meets the *scrotum*, or scrotal sac, at the penoscrotal junction. Located posterior to the penis and anterior to the anus, the scrotum is an extra-abdominal pouch that consists of a thin layer of skin overlying a tighter, muscle-like layer. This muscle-like layer, in turn, overlies the tunica vaginalis, a serous membrane that covers the internal scrotal cavity.

Canals and rings

Internally, a *septum* divides the scrotum into two sacs, which each contain a *testis*, an *epididymis*, and a *sper-*

matic cord. The spermatic cord is a connective tissue sheath that encases autonomic nerve fibers, blood vessels, lymph vessels, and the *vas deferens* (also called *ductus deferens*).

The spermatic cord travels from the testis through the inguinal canal, exiting the scrotum through the external inguinal ring and entering the abdominal cavity through the internal inguinal ring. The inguinal canal lies between the two rings.

Loads of nodes

Lymph nodes from the penis, scrotal surface, and anus drain into the *inguinal* lymph nodes. Lymph nodes from the testes drain into the lateral aortic and pre-aortic lymph nodes in the abdomen.

Testes

The testes are enveloped in two layers of connective tissue: the *tunica vaginalis* (outer layer) and the *tunica albuginea* (inner layer). Extensions of the tunica albuginea separate the testis into lobules. Each lobule contains one to four *seminiferous tubules*, small tubes where spermatogenesis takes place.

Climate control

Spermatozoa development requires a temperature lower than that of the rest of the body. The *dartos muscle,* a smooth muscle in the superficial fascia, causes scrotal skin to wrinkle, which helps regulate temperature. The *cremaster muscle,* rising from the internal oblique muscle, helps to govern temperature by elevating the testes.

Duct system

The male reproductive *duct system,* consisting of the epididymis, vas deferens, and urethra, conveys sperm from the testes to the ejaculatory ducts near the bladder.

Swimmer storage

The *epididymis* is a coiled tube located superior to and along the posterior border of the testis. During ejaculation, smooth muscle in the epididymis contracts, ejecting spermatozoa into the vas deferens.

An important function of the scrotum is to keep the testes cooler than the rest of the body.

Descending tunnel

The *vas deferens* leads from the testes to the abdominal cavity and extends upward through the *inguinal canal*, arches over the urethra, and descends behind the bladder. Its enlarged portion, called the *ampulla*, merges with the duct of the seminal vesicle to form the short ejaculatory duct. After passing through the prostate gland, the vas deferens joins with the urethra.

Tube to the outside

A small tube leading from the floor of the bladder to the exterior, the *urethra* consists of three parts:
• *prostatic urethra*, surrounded by the prostate gland, which drains the bladder
• *membranous urethra*, which passes through the urogenital diaphragm
• *spongy urethra*, which makes up about 75% of the entire urethra.

During ejaculation, smooth muscle in the epididymis contracts, sending spermatozoa into the vas deferens.

Accessory reproductive glands

The accessory glands, which produce most of the semen, include the *seminal vesicles, bulbourethral glands (Cowper's glands),* and *prostate gland.*

The seminal vesicles are paired sacs at the base of the bladder. The bulbourethral glands, also paired, are located inferior to the prostate.

Size of a walnut

The walnut-sized prostate gland lies under the bladder and surrounds the urethra. It consists of three lobes: the left and right lateral lobes and the median lobe.

Improving the odds

The prostate continuously secretes prostatic fluid, a thin, milky, alkaline fluid. During sexual activity, prostatic fluid adds volume to the semen. The fluid enhances sperm motility and may improve the odds for conception by neutralizing the acidity of the urethra and of the woman's vagina.

Slightly alkaline

Semen is a viscous, white secretion with a slightly alkaline pH (7.8 to 8) and consists of spermatozoa and accessory gland secretions. The seminal vesicles produce roughly 60% of the fluid portion of the semen, while the prostate gland produces about 30%. A viscid fluid secreted by the bulbourethral glands also becomes part of the semen.

Memory jogger

To remember the meaning of spermatogenesis, keep in mind that genesis means "beginning" or "new." Therefore, spermatogenesis means beginning of new sperm.

Spermatogenesis

Sperm formation, or *spermatogenesis*, begins when a male reaches *puberty* and normally continues throughout life.

Divide and conquer

Spermatogenesis occurs in four stages:
• In the first stage, the primary germinal epithelial cells, called *spermatogonia*, grow and develop into primary *spermatocytes*. Both spermatogonia and primary spermatocytes contain 46 chromosomes, consisting of 44 *autosomes* and the two sex chromosomes, X and Y.
• Next, primary spermatocytes divide to form secondary spermatocytes. No new chromosomes are formed in this stage; the pairs only divide. Each secondary spermatocyte contains half the number of autosomes, 22; one secondary spermatocyte contains an X chromosome, the other, a Y chromosome.
• In the third stage, each secondary spermatocyte divides again to form *spermatids* (also called *spermatoblasts*).
• Finally, the spermatids undergo a series of structural changes that transform them into mature *spermatozoa*, or sperm. Each spermatozoa has a head, neck, body, and tail. The head contains the *nucleus*, the tail, a large amount of *adenosine triphosphate*, which provides energy for sperm *motility*.

Queuing up

Newly mature sperm pass from the seminiferous tubules through the *vasa recta* into the epididymis. Only a small number of sperm can be stored in the epididymis. Most of them move into the vas deferens, where they're stored until sexual stimulation triggers emission.

Check the expiration date?

Sperm cells retain their potency in storage for many weeks. After ejaculation, sperm survive for 24 to 72 hours at body temperature.

Hormonal control and sexual development

Androgens (male sex hormones) are produced in the testes and the adrenal glands. Androgens are responsible for the development of male sex organs and secondary sex characteristics. Major androgens include:

- testosterone
- luteinizing hormone (LH)
- follicle-stimulating hormone (FSH).

The number of sperm and their motility affect fertility. A low sperm count (less than 20 million per milliliter of ejaculated semen) may cause infertility.

The captain of the team

Leydig's cells, located in the testes between the seminiferous tubules, secrete *testosterone*, the most significant male sex hormone.

Testosterone is responsible for the development and maintenance of male sex organs and secondary sex characteristics, such as facial hair and vocal cord thickness. Testosterone is also required for spermatogenesis.

Calling the plays

Testosterone secretion begins approximately 2 months after conception, when the release of chorionic gonadotropins from the placenta stimulates Leydig's cells in the male fetus. The presence of testosterone directly affects sexual differentiation in the fetus. With testosterone, fetal genitalia develop into a penis, scrotum, and testes; without testosterone, genitalia develop into a clitoris, vagina, and other female organs.

During the last 2 months of gestation, testosterone normally causes the testes to descend into the scrotum. If the testes don't descend after birth, exogenous testosterone may correct the problem.

Other key players

Other hormones also affect male sexuality. Two of these, *LH* — also called *interstitial cell-stimulating hormone* — and *FSH*, directly affect secretion of testosterone.

Time to grow

During early childhood, a boy doesn't secrete gonadotropins and thus has little circulating testosterone. Secretion of gonadotropins from the pituitary gland, which usually occurs between ages 11 and 14, marks the onset of *puberty*. These pituitary gonadotropins stimulate testis functioning as well as testosterone secretion.

From boy to man

During puberty, the penis and testes enlarge and the male reaches full adult sexual and reproductive capability. Puberty also marks the development of male secondary sexual characteristics: distinct body hair distribution, skin changes (such as increased secretion by sweat and sebaceous glands), deepening of the voice (from laryngeal enlargement), increased musculoskeletal development, and other intracellular and extracellular changes.

Reaching the plateau

After a male achieves full physical maturity, usually by age 20, sexual and reproductive function remain fairly consistent throughout life.

Subtle changes

With aging, a man may experience subtle changes in sexual function, but he doesn't lose the ability to reproduce. For example, an elderly man may require more time to achieve an erection, experience less firm erections, and have reduced ejaculatory volume. After ejaculation, he may take longer to regain an erection.

I've reached puberty. Now can I borrow the car?

Female reproductive system

Unlike the male reproductive system, the female system is largely internal, housed within the pelvic cavity.

External genitalia

The vulva contains the external female genitalia, visible on inspection. It includes the *mons pubis, labia majora, labia minora, clitoris,* and adjacent structures. (See *Structures of the female reproductive system*, pages 284 and 285.)

At the bottom

The *mons pubis* is a rounded cushion of fatty and connective tissue covered by skin and coarse, curly hair in a triangular pattern over the symphysis pubis (the joint formed by the union of the pubic bones anteriorly).

Around the outside, part 1

The *labia majora* are two raised folds of adipose and connective tissue that border the vulva on either side, extending from the mons pubis to the perineum. After *menarche* (onset of the first menstrual period), the outer surface of the labia is covered with pubic hair. The inner surface is pink and moist.

Around the outside, part 2

The *labia minora* are two moist folds of mucosal tissue, dark pink to red in color, that lie within and alongside the labia majora. Each upper section divides into an upper and lower lamella. The two upper lamellae join to form the *prepuce,* a hoodlike covering over the clitoris. The two lower lamellae form the *frenulum,* the posterior portion of the clitoris.

The lower labial sections taper down and back from the clitoris to the perineum, where they join to form the *fourchette,* a thin tissue fold along the anterior edge of the perineum.

Minor in name only

The labia minora contain sebaceous glands, which secrete a lubricant that also acts as a bactericide. Like the labia majora, they're rich in blood vessels and nerve endings, making them highly responsive to stimulation. They swell in response to sexual stimulation, a reaction that triggers other changes that prepare the genitalia for coitus.

Specially sensitive

The *clitoris* is the small, protuberant organ just beneath the arch of the mons pubis. It contains erectile tissue, venous cavernous spaces, and specialized sensory corpuscles, which are stimulated during sexual activity.

Featuring glands

The *vestibule* is an oval area bounded anteriorly by the clitoris, laterally by the labia minora, and posteriorly by the fourchette. The mucus-producing *Skene's glands* are found on both sides of the urethral opening. Openings of the two mucus-producing *Bartholin's glands* are located

The labia are highly vascular and have many nerve endings — making them sensitive to pain, pressure, touch, sexual stimulation, and temperature extremes.

Zoom in

Structures of the female reproductive system

The female reproductive system includes the vagina, cervix, uterus, fallopian tubes, ovaries, and other structures.

View of external genitalia in lithotomy position

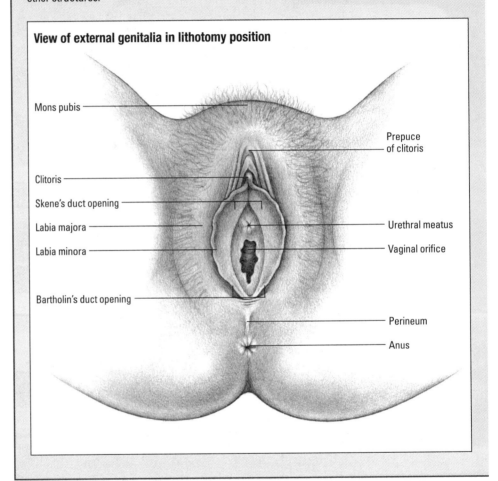

Mons pubis

Clitoris

Skene's duct opening

Labia majora

Labia minora

Bartholin's duct opening

Prepuce of clitoris

Urethral meatus

Vaginal orifice

Perineum

Anus

laterally and posteriorly on either side of the inner vaginal orifice.

The *urethral meatus* is the slitlike opening below the clitoris through which urine leaves the body. In the center of the vestibule is the *vaginal orifice*. It may be completely or partially covered by the *hymen*, a tissue membrane.

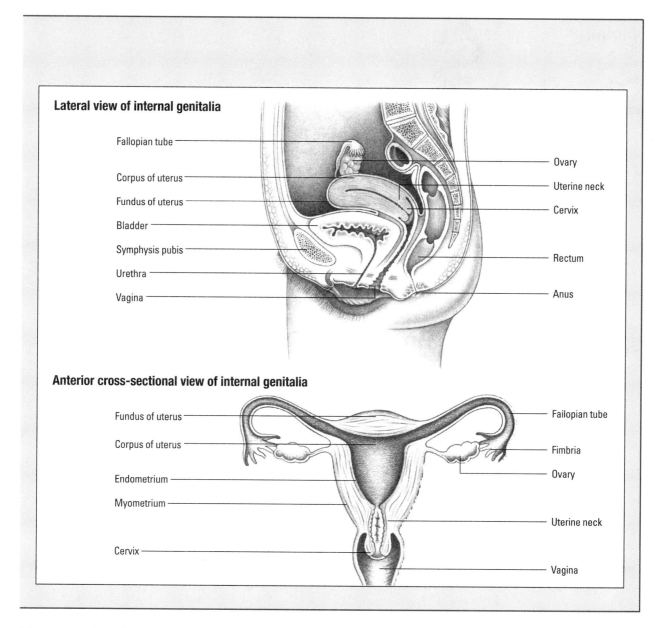

Lateral view of internal genitalia

Fallopian tube

Corpus of uterus

Fundus of uterus

Bladder

Symphysis pubis

Urethra

Vagina

Ovary

Uterine neck

Cervix

Rectum

Anus

Anterior cross-sectional view of internal genitalia

Fundus of uterus

Corpus of uterus

Endometrium

Myometrium

Cervix

Failopian tube

Fimbria

Ovary

Uterine neck

Vagina

Not too simple

Located between the lower vagina and the anal canal, the *perineum* is a complex structure of muscles, blood vessels, fascia, nerves, and lymphatics.

Vagina

The female internal genitalia, beginning with the vagina, are specialized organs the main function of which is reproduction.

Three layers...

The *vagina*, a highly elastic muscular tube, is located between the urethra and the rectum. The vaginal wall has three tissue layers: epithelial tissue, loose connective tissue, and muscle tissue. The *uterine cervix* connects the uterus to the vaginal vault. Four *fornices*, recesses in the vaginal wall, surround the cervix.

...three functions

The vagina has three main functions:
• to accommodate the penis during coitus
• to channel blood discharged from the uterus during menstruation
• to serve as the birth canal during childbirth.

Supplied separately

The upper, middle, and lower vaginal sections have separate blood supplies. Branches of the uterine arteries supply blood to the upper vagina; the *inferior vesical arteries* supply blood to the middle vagina; and the hemorrhoidal and *internal pudendal arteries* feed into the lower vagina.

Blood returns through a vast venous *plexus* to the hemorrhoidal, pudendal, and uterine veins and then to the *hypogastric* veins. This plexus merges with the *vertebral venous plexus*.

Cervix

The cervix projects into the upper portion of the vagina. The lower cervical opening is the *external os*, the upper opening is the *internal os*.

Permanent alterations

Childbirth permanently alters the cervix. In a female who hasn't delivered a child, the external os is a round opening about 3 mm in diameter; after the first childbirth, it becomes a small transverse slit with irregular edges.

Uterus

The *uterus* is a small, firm, pear-shaped, muscular organ situated between the bladder and rectum. It typically lies at almost a 90-degree angle to the vagina. The mucous membrane lining of the uterus is called the *endometrium*, and the muscular layer of the uterus is called the *myometrium*.

Fundamental fundus

During pregnancy, the elastic, upper portion of the uterus, called the *fundus*, accommodates most of the growing fetus until term. The uterine neck joins the fundus to the cervix, the uterine part extending into the vagina. The fundus and neck make up the *corpus*, the main uterine body.

Over a woman's lifetime, the size of her uterus and cervix changes.

Fallopian tubes

Two *fallopian tubes* attach to the uterus at the upper angles of the fundus. These narrow cylinders of muscle fibers are the site of fertilization.

Riding the wave

The curved portion of the fallopian tube, called the *ampulla*, ends in the funnel-shaped *infundibulum*. Fingerlike projections in the infundibulum, called *fimbriae*, move in waves that sweep the mature ovum from the ovary into the fallopian tube.

Fingerlike projections called fimbriae move in waves, sweeping the ovum from the ovary to the fallopian tube.

Ovaries

The *ovaries* are located on either side of the uterus. The size, shape, and position of the ovaries vary with age. Round, smooth, and pink at birth, they grow larger, flatten, and turn grayish by puberty. During the childbearing years, they take on an almond shape and a rough, pitted surface; after menopause, they shrink and turn white.

50,000 follicles

The ovaries' main function is to produce mature ova. At birth, each ovary contains approximately 50,000 *graafian follicles*. During the childbearing years, one graafian follicle produces a mature ovum during the

first half of each menstrual cycle. As the ovum matures, the follicle ruptures and the ovum is swept into the fallopian tube.

The ovaries also produce estrogen and progesterone as well as a small amount of androgens.

Mammary glands

The mammary glands, located in the breast, are specialized accessory glands that secrete milk. Although present in both sexes, they typically function only in the female.

15 to 25 lobes

Each mammary gland contains 15 to 25 lobes separated by fibrous connective tissue and fat. Within the lobes are clustered acini — tiny, saclike duct terminals that secrete milk during lactation.

The ducts draining the lobules converge to form excretory (*lactiferous*) ducts and sinuses (*ampullae*), which store milk during lactation. These ducts drain onto the nipple surface through 15 to 20 openings. (See *The female breast.*)

Both males and females have mammary glands — but they function only in the female.

Hormonal function and the menstrual cycle

Like the male body, the female body changes as it ages in response to hormonal control. When a female reaches the age of menstruation, the hypothalamus, ovaries, and pituitary gland secrete hormones — *estrogen, progesterone, FSH,* and *LH* — that affect the buildup and shedding of the endometrium during the menstrual cycle. (See *Events in the female reproductive cycle,* pages 290 and 291.)

A sometimes shocking development

During adolescence, the release of hormones causes a rapid increase in physical growth and spurs the development of secondary sex characteristics. This growth spurt begins at approximately age 11 and continues until early adolescence, or about 3 years later.

Irregularity of the menstrual cycle is common during this time because of failure of the female to ovulate. With menarche, the uterine body flexes on the cervix and the ovaries are situated in the pelvic cavity.

(Text continues on page 292.)

Zoom in

The female breast

The breasts are located on either side of the anterior chest wall over the greater pectoral and the anterior serratus muscles. Within the areola, the pigmented area in the center of the breast, lies the nipple. Erectile tissue in the nipple responds to cold, friction, and sexual stimulation.

Support and separate

Each breast is composed of glandular, fibrous, and adipose tissue. Glandular tissue contains 15 to 20 lobes made up of clustered *acini,* tiny saclike duct terminals that secrete milk. Fibrous *Cooper's ligaments* support the breasts; *adipose tissue* separates the two breasts.

Produce and drain

Acini draw the ingredients to make milk from the blood in surrounding capillaries.

Sebaceous glands on the areolar surface, called *Montgomery's tubercles*, produce *sebum,* which lubricates the areola and nipple during breast-feeding.

Lateral cross section

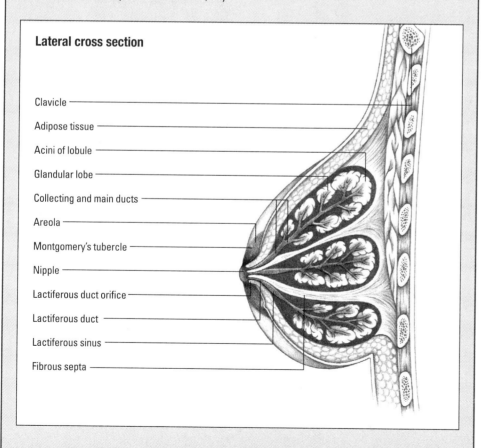

Clavicle

Adipose tissue

Acini of lobule

Glandular lobe

Collecting and main ducts

Areola

Montgomery's tubercle

Nipple

Lactiferous duct orifice

Lactiferous duct

Lactiferous sinus

Fibrous septa

Now I get it!

Events in the female reproductive cycle

The female reproductive cycle usually lasts 28 days. During this cycle, three major types of changes occur simultaneously: ovulatory, hormonal, and endometrial (involving the lining [endometrium] of the uterus).

Ovulatory
- Ovulatory changes begin on the 1st day of the menstrual cycle, which usually lasts 5 days.
- As the cycle begins, low estrogen and progesterone levels in the bloodstream stimulate the hypothalamus to secrete gonadotropin-stimulating hormone (Gn-RH). In turn, Gn-RH stimulates the anterior pituitary gland to secrete follicle-stimulating hormone (FSH) and luteinizing hormone (LH).
- Follicle development within the ovary (in the follicular stage) is spurred by increased levels of FSH and, to a lesser extent, LH.
- When the follicle matures, a spike in LH level occurs, causing the follicle to rupture and release the ovum, thus initiating ovulation.
- After ovulation (in the luteal stage), the collapsed follicle forms the corpus luteum, which (if fertilization doesn't occur) degenerates.

Hormonal
- During the follicular phase of the ovarian cycle, the increasing FSH and LH levels that stimulate follicle growth also stimulate increased secretion of estrogen.
- Estrogen secretion peaks just before ovulation. This peak sets in motion the spike in LH levels, which causes ovulation.
- After ovulation, estrogen levels decline rapidly. In the luteal phase of the ovarian cycle, the corpus luteum is formed and beings to release progesterone and estrogen.
- As the corpus luteum degenerates, levels of both of these ovarian hormones decline.

Endometrial
- The endometrium is receptive to implantation of an embryo for only a short time in the reproductive cycle. Thus, it's no accident that its most receptive stage occurs about 7 days after the ovarian cycle's release of an ovum — just in time to receive a fertilized ovum.
- In the first 5 days of the reproductive cycle, the endometrium sheds its functional layer, leaving the basal layer (the deepest layer) intact. Menstrual flow consists of this detached layer and accompanying blood from the detachment process.
- The endometrium begins regenerating its functional layer at about day 6 (the proliferative stage), spurred by rising estrogen levels.
- After ovulation (about day 14), increased progesterone secretion stimulates conversion of the functional layer into a secretory mucosa (secretory stage), which is more receptive to implantation of the fertilized ovum.
- If implantation doesn't occur, the corpus luteum degenerates, progesterone levels drop, and the endometrium again sheds its functional layer.

There are three major types of changes during the reproductive cycle: ovulatory, hormonal, and endometrial.

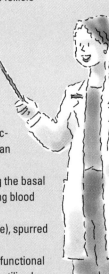

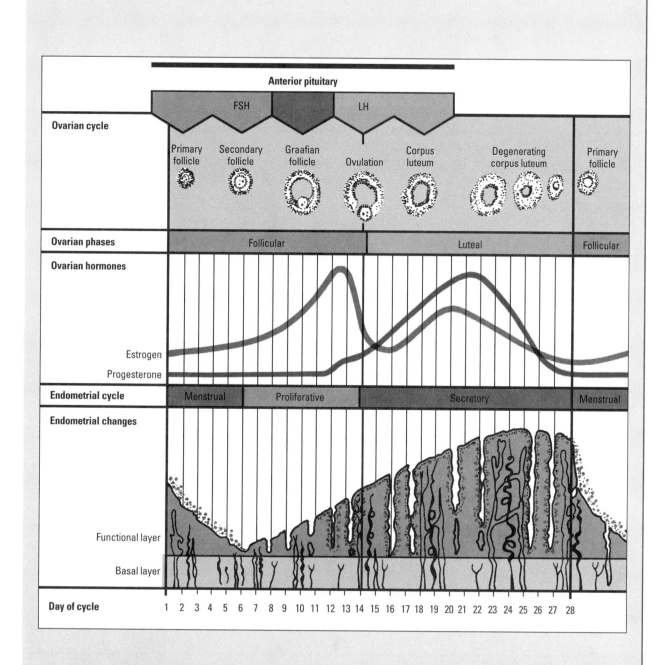

A monthly thing

The menstrual cycle is a complex process that involves the reproductive and endocrine systems. The cycle averages 28 days in length.

Hormone slow-down

In contrast to the slowly declining hormones of the aging male, the female's hormones decline rapidly. The stage of menopause includes the preceding 1 to 2 years of declining ovarian function, shown by irregular menses that gradually become further apart and produce a lighter flow. The ovaries stop producing progesterone and estrogen.

Supply exhausted

Cessation of menses usually occurs between ages 40 and 55. Although the pituitary gland still releases FSH and LH, the body has exhausted the supply of ovarian follicles that respond to these hormones, and menstruation no longer occurs.

1-year marker

A woman is considered to have reached menopause after menses are absent for 1 year. Previous to menopause, a woman experiences several transitional years (called the climacteric years), during which several physiologic changes that lead to menopause occur.

The menstrual cycle may range from 22 to 34 days, although the typical cycle lasts 28 days.

Quick quiz

1. Spermatogenesis is:
   A. growth and development of sperm into primary spermatocytes.
   B. division of spermatocytes to form secondary spermatocytes.
   C. entire process of sperm formation.

Answer: C. Spermatogenesis refers to the entire process of sperm formation, from the development of primary spermatocytes to the formation of fully functional spermatozoa.

2. The primary function of the scrotum is to:
   A. provide storage for newly developed sperm.
   B. maintain a cool temperature for the testes.
   C. deposit sperm in the female reproductive tract.

Answer: B. The function of the scrotum is to maintain a cool temperature for the testes, which is necessary for spermatozoa formation.

3. The main function of the ovaries is to:
   A. secrete hormones that affect the buildup and shedding of the endometrium during the menstrual cycle.
   B. accommodate a growing fetus during pregnancy.
   C. produce mature ova.

Answer: C. The main function of the ovaries is to produce mature ova.

4. The corpus luteum forms and degenerates in which stage of the female reproductive cycle?
   A. Luteal
   B. Follicular
   C. Proliferative

Answer: A. The corpus luteum forms and degenerates in the luteal stage, a phase of the ovarian cycle of reproduction.

5.   The four hormones involved in the menstrual cycle are:

A.   luteinizing hormone (LH), progesterone, estrogen, and testosterone.

B.   estrogen, follicle-stimulating hormone (FSH), LH, and androgens.

C.   estrogen, progesterone, LH, and FSH.

Answer: C. The four hormones involved in the menstrual cycle are estrogen, progesterone, LH, and FSH.

Scoring

☆☆☆   If you answered all five questions correctly, congrats! You've hit a growth spurt in your knowledge.

☆☆   If you answered three or four questions correctly, good for you! You've got a firm grasp of this complicated system.

☆   If you answered fewer than three questions correctly, keep trying! Like the reproductive system, your knowledge may just need some time to develop.

Reproduction and lactation

Just the facts

In this chapter, you'll learn:
♦ how fertilization occurs
♦ how the embryo and fetus develop
♦ stages of labor
♦ process of lactation.

Fertilization

Production of a new human being begins with *fertilization*, the union of a *spermatozoon* and an *ovum* to form a single cell. After fertilization occurs, dramatic changes begin inside a woman's body. The cells of the fertilized ovum begin dividing as the ovum travels to the *uterine cavity*, where it implants in the uterine lining. (See *How fertilization occurs*, page 296.)

One in a million

For fertilization to take place, however, the spermatozoon must first reach the ovum.

Although a single ejaculation deposits several hundred million spermatozoa, many are destroyed by acidic vaginal secretions. The only spermatozoa that survive are those that enter the *cervical canal*, where *cervical mucus* protects them.

Timing is everything

The ability of spermatozoa to penetrate the cervical mucus depends on the phase of the menstrual cycle at the time of transit.

Now I get it!

How fertilization occurs

Fertilization begins when the *spermatozoon* is activated upon contact with the *ovum*.

The spermatozoon has a covering called the *acrosome* that develops small perforations through which it releases enzymes necessary for the sperm to penetrate the protective layers of the ovum before fertilization.

The spermatozoon then penetrates the *zona pellucida* (the inner membrane of the ovum). This triggers the ovum's second *meiotic* division (following meiosis), making the zona pellucida impenetrable to other spermatozoa.

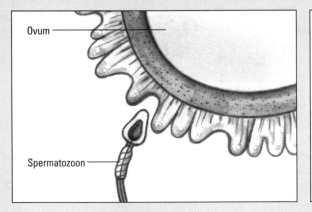

Ovum

Spermatozoon

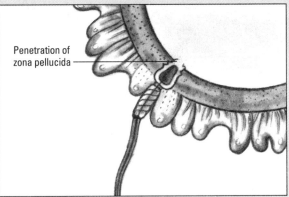

Penetration of
zona pellucida

After the spermatozoon penetrates the ovum, its nucleus is released into the ovum, its tail degenerates, and its head enlarges and fuses with the nucleus of the ovum. This fusion provides the fertilized ovum, called a *zygote*, with 46 chromosomes.

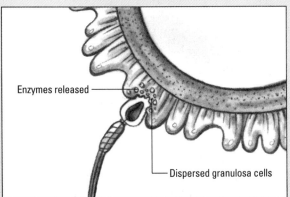

Enzymes released

Dispersed granulosa cells

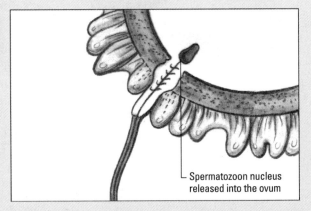

Spermatozoon nucleus
released into the ovum

Early in the cycle, estrogen and progesterone levels cause the mucus to thicken, making it more difficult for spermatozoa to pass through the cervix. During midcycle, however, when the mucus is relatively thin, spermatozoa can pass readily through the cervix. Later in the cycle, the cervical mucus thickens again, hindering spermatozoa passage.

Help along the way

Spermatozoa travel through the female reproductive tract at a rate of several millimeters per hour by means of flagellar (whiplike movements of the tail) movements.

After spermatozoa pass through the cervical mucus, however, the female reproductive system "assists" them on their journey with rhythmic contractions of the uterus that help the spermatozoa to penetrate the fallopian tubes. Spermatozoa are typically viable (able to fertilize the ovum) for up to 2 days after ejaculation; however, they can survive in the reproductive tract for up to 4 days.

Success becomes a zygote

Before a spermatozoon can penetrate the ovum, it must disperse the *granulosa* cells and penetrate the *zona pellucida*, the thick, transparent layer surrounding the incompletely developed ovum. Enzymes in the *acrosome* (head cap) of the spermatozoon permit this penetration. After penetration, the ovum completes its second meiotic division, and the zona pellucida prevents penetration by other spermatozoa.

The spermatozoon's head then fuses with the ovum nucleus, creating a cell nucleus with 46 chromosomes. The fertilized ovum is called a zygote.

Congratulations! You're a zygote.

Pregnancy

Pregnancy starts with fertilization and ends with childbirth; on average, its duration is 38 weeks. During this period (called *gestation*), the zygote divides as it passes through the fallopian tube and attaches to the uterine lining by implantation. A complex sequence of *preembryonic*, *embryonic*, and *fetal* development transforms the zygote into a full-term fetus.

Making predictions

Because the uterus grows throughout pregnancy, uterine size serves as a rough estimate of the duration of pregnancy. The actual fertilization date is rarely known, so the woman's expected delivery date is typically calculated from the beginning of her last menstrual period.

Now I get it!

Preembryonic development

The preembryonic phase lasts from conception until approximately the end of the 3rd week of development.

Zygote formation. . .
As the fertilized ovum advances through the fallopian tube toward the uterus, it undergoes *mitotic division*, forming daughter cells, initially called *blastomeres*, that each contain the same number of chromosomes as the parent cell. The first cell division ends about 30 hours after fertilization; subsequent divisions occur rapidly.

The *zygote*, as it is now called, develops into a small mass of cells called a *morula*, which reaches the uterus at about the 3rd day after fertilization. Fluid that amasses in the center of the morula forms a central cavity.

. . . into blastocyst
The structure is now called a *blastocyst*. Its cells differentiate into one of two forms: the *trophoblast,* which develops into fetal membranes and contributes to placenta formation (these cells are called *blastocytes*); or the inner cell mass, a discrete cell cluster enclosed within the trophoblast, which will form the embryo (late *blastocyst*).

Getting attached: Blastocyst and endometrium
During the next phase, the blastocyst stays within the zona pellucida, unattached to the uterus. The zona pellucida degenerates and, by the end of the 1st week after fertilization, the blastocyst attaches to the *endometrium*. The part of the blastocyst adjacent to the inner cell mass is the first part to become attached.

The trophoblast, in contact with the endometrial lining, proliferates and invades the underlying endometrium by separating and dissolving endometrial cells.

Letting it all sink in
During the next week, the invading blastocyst sinks below the surface of the endometrium. The site of penetration seals, restoring the continuity of the endometrial surface.

Slide rule

A tool for calculating delivery dates is Naegele's rule. If you know the 1st day of the last menstrual cycle, simply count back 3 months from that date and then add 7 days.

Stages of fetal development

During pregnancy, the fetus undergoes three major stages of development:
- preembryonic period (fertilization to week 3)
- embryonic period (weeks 4 through 7)
- fetal period (week 8 through birth).

The first 3 weeks

The preembryonic phase starts with ovum fertilization and lasts 3 weeks. As the zygote passes through the fallopian tube, it undergoes a series of *mitotic divisions*, or *cleavage*. (See *Preembryonic development*.)

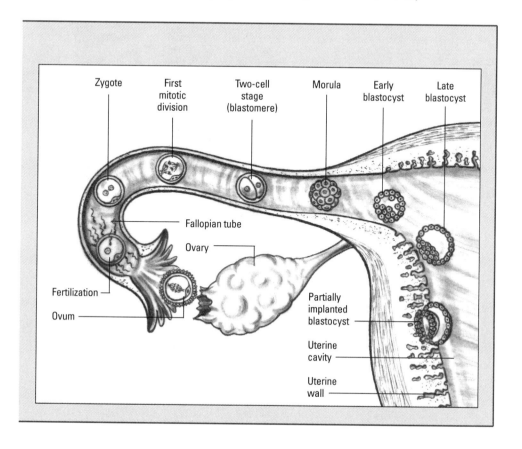

Zygote   First mitotic division   Two-cell stage (blastomere)   Morula   Early blastocyst   Late blastocyst

Fallopian tube

Ovary

Fertilization

Ovum

Partially implanted blastocyst

Uterine cavity

Uterine wall

Now I get it!

Embryonic development

Each of the three germ layers — *ectoderm*, *mesoderm*, and *endoderm* — forms specific tissues and organs in the developing embryo.

Ecto becomes epidermis

The ectoderm, the outermost layer, develops into the:

- epidermis
- nervous system
- pituitary gland
- tooth enamel
- salivary glands
- optic lens
- lining of lower portion of anal canal
- hair.

Meso makes musculature

The mesoderm, the middle layer, develops into:

- connective and supporting tissue
- the blood and vascular system
- musculature
- teeth (except enamel)
- the mesothelial lining of pericardial, pleural, and peritoneal cavities
- the kidneys and ureters.

Endo equals epithelial

The endoderm, the innermost layer, becomes the epithelial lining of the:

- pharynx and trachea
- auditory canal
- alimentary canal
- liver
- pancreas
- bladder and urethra
- prostate.

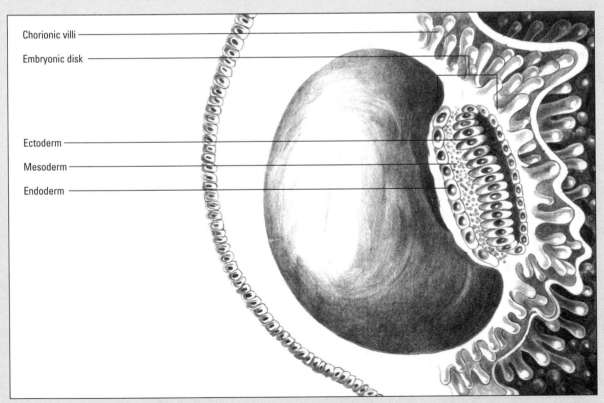

Chorionic villi

Embryonic disk

Ectoderm

Mesoderm

Endoderm

The next month

During the embryonic period (the 4th through the 7th week of gestation), the developing zygote starts to take on a human shape and is now called an *embryo.* Each germ layer — the *ectoderm, mesoderm,* and *endoderm* — eventually forms specific tissues in the embryo. (See *Embryonic development.*)

A cautionary tale

The organ systems form during the embryonic period. During this time, the embryo is particularly vulnerable to injury by maternal drug use, certain maternal infections, and other factors.

The rest of the way

During the *fetal* stage of development, which extends from the 8th week until birth, the maturing fetus enlarges and grows heavier. (See *From embryo to fetus,* page 302.)

Two unusual features appear during this stage:

- the fetus' head is disproportionately large compared with its body (This feature changes after birth as the infant grows.)
- the fetus lacks subcutaneous fat. (Fat starts to accumulate shortly after birth.)

The fetus isn't the only thing changing during pregnancy — the reproductive system undergoes changes as well.

Structural changes in the ovaries and uterus

Pregnancy changes the usual development of the *corpus luteum* and results in development of the following structures:
- decidua
- amniotic sac and fluid
- yolk sac
- placenta.

Corpus luteum

Normal functioning of the corpus luteum requires continual stimulation by *luteinizing hormone (LH).* Progesterone produced by the corpus luteum suppresses LH release by the pituitary gland. If pregnancy occurs, the corpus luteum continues to produce progesterone until

Now I get it!

From embryo to fetus

Significant growth and development take place within the first 3 months following conception, as the embryo develops into a fetus that nearly resembles a full-term newborn.

Month 1

At the end of the 1st month, the embryo has a definite form. The head, the trunk, and the tiny buds that will become the arms and legs are discernible. The cardiovascular system has begun to function, and the umbilical cord is visible in its most primitive form.

Month 2

During the 2nd month, the embryo — called a *fetus* from the 8th week — grows to 1" (2.5 cm) in length and weighs 1/30 oz (1 g). The head and facial features develop as the eyes, ears, nose, lips, tongue, and tooth buds form. The arms and legs also take shape. Although the gender of the fetus isn't yet discernible, all external genitalia are present. Cardiovascular function is complete, and the umbilical cord has a definite form. At the end of the 2nd month, the fetus resembles a full-term newborn except for size.

Month 3

During the 3rd month, the fetus grows to 3" (7.5 cm) in length and weighs 1 oz (28 g). Teeth and bones begin to appear, and the kidneys start to function. The fetus opens its mouth to swallow, grasps with its fully developed hands, and prepares for breathing by inhaling and exhaling (although its lungs aren't functioning). At the end of the first *trimester* (the 3-month periods into which pregnancy is divided), its gender is distinguishable.

Months 3 to 9

Over the remaining 6 months, fetal growth continues as internal and external structures develop at a rapid rate. In the third trimester, the fetus stores the fats and minerals it will need to live outside the womb. At birth, the average full-term fetus measures 20" (51 cm) and weighs 7 to 7½ lb (3.2 to 3.4 kg).

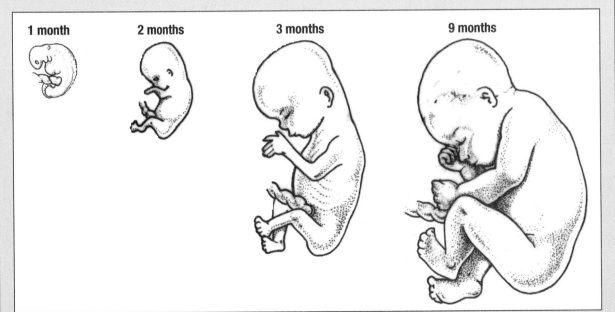

1 month   **2 months**   **3 months**   **9 months**

the placenta takes over. Otherwise, the corpus luteum atrophies 3 days before menstrual flow begins.

Hormone soup

With age, the corpus luteum grows less responsive to LH. For this reason, the mature corpus luteum degenerates unless stimulated by progressively increasing amounts of LH.

Pregnancy stimulates the placental tissue to secrete large amounts of *human chorionic gonadotropin (HCG)*, which resembles LH and *follicle-stimulating hormone (FSH)*, also produced by the pituitary gland. HCG prevents corpus luteum degeneration, stimulating the corpus luteum to produce large amounts of estrogen and progesterone.

Early detection through HCG

The corpus luteum, stimulated by the hormone HCG, produces the estrogen and progesterone needed to maintain the pregnancy during the first 3 months. HCG can be detected as early as 9 days after fertilization and can provide confirmation of pregnancy even before the woman has missed her first menstrual period.

The HCG level gradually increases during this time, peaks at about 10 weeks' gestation, and then gradually declines.

Decidua

The *decidua* is an endometrial lining of the uterus that undergoes the hormone-induced changes of pregnancy. Decidual cells secrete the following three substances:
• the hormone *prolactin*, which promotes lactation
• a peptide hormone, *relaxin*, which induces relaxation of the connective tissue of the symphysis pubis and pelvic ligaments; it also promotes cervical dilation
• a potent hormone-like fatty acid, *prostaglandin*, which mediates several physiologic functions.

(See *Development of the decidua and fetal membranes*, page 304.)

Amniotic sac and fluid

The *amniotic sac*, enclosed within the chorion, gradually increases in size and surrounds the embryo. As it enlarges, the amniotic sac expands into the chorionic cavi-

Now I get it!

Development of the decidua and fetal membranes

Specialized tissues support, protect, and nurture the embryo and fetus throughout its development. Among these tissues, the decidua and fetal membranes begin to develop shortly after conception.

Nesting place

During pregnancy, the endometrial lining is called the *decidua*. It provides a nesting place for the developing ovum and has some endocrine functions.

Based primarily on its position relative to the embryo, the decidua may be known as the *decidua basalis,* which lies beneath the *chorionic vesicle* (see below), the *decidua capsularis,* which stretches over the vesicle, or the *decidua parietalis,* which lines the remainder of the endometrial cavity.

Network of blood vessels

The *chorion* is a membrane that forms the outer wall of the blastocyst. Vascular projections, called *chorionic villi,* arise from its periphery. As the *chorionic vesicle* enlarges, villi arising from the superficial portion of the chorion, called the *chorion laeve,* atrophy, leaving this surface smooth. Villi arising from the deeper part of the chorion, called the *chorion frondosum,* proliferate, projecting into the large blood vessels within the decidua basalis through which the maternal blood flows.

Blood vessels form within the villi as they grow, and they become connected with blood vessels that form in the chorion, body stalk, and within the body of the embryo. Blood begins to flow through this developing network of vessels as soon as the embryo's heart starts to beat.

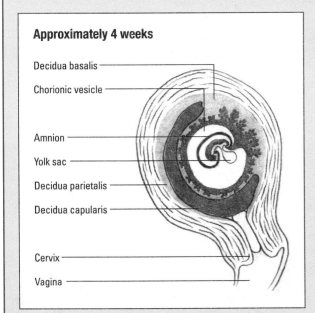

Approximately 4 weeks

Decidua basalis
Chorionic vesicle
Amnion
Yolk sac
Decidua parietalis
Decidua capularis
Cervix
Vagina

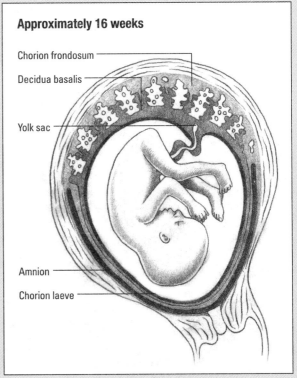

Approximately 16 weeks

Chorion frondosum
Decidua basalis
Yolk sac
Amnion
Chorion laeve

ty, eventually filling the cavity and fusing with the chorion by the 8th week of gestation.

A warm, protective sea

The amniotic sac and amniotic fluid serve the fetus in two important ways, one during gestation and the other during delivery. During gestation, the fluid gives the fetus a buoyant, temperature-controlled environment. Later, amniotic fluid serves as a fluid wedge that helps open the cervix during birth.

Some from the mother, some from the fetus

Early in pregnancy, amniotic fluid comes chiefly from three sources:
• fluid filtering into the amniotic sac from maternal blood as it passes through the uterus
• fluid filtering into the sac from fetal blood passing through the placenta
• fluid diffusing into the amniotic sac from the fetal skin and respiratory tract.

Later in pregnancy, when the fetal kidneys begin to function, the fetus urinates into the amniotic fluid. Fetal urine then becomes the major source of amniotic fluid.

Every sea has its tides

Production of amniotic fluid from maternal and fetal sources balances amniotic fluid that is lost through the fetal GI tract. Normally, the fetus swallows up to several hundred milliliters of amniotic fluid each day. The fluid is absorbed into the fetal circulation from the fetal GI tract; some is transferred from the fetal circulation to the maternal circulation and excreted in maternal urine.

Yolk sac

The *yolk sac* forms next to the endoderm of the germ disk; a portion of it is incorporated in the developing embryo and forms the GI tract. Another portion of the sac develops into primitive germ cells, which travel to the developing gonads and eventually form *oocytes* (the precursor of the ovum) or *spermatocytes* (the precursor of the spermatozoon) after gender has been determined.

Amniotic fluid protects the fetus during pregnancy — and later helps open the cervix for delivery.

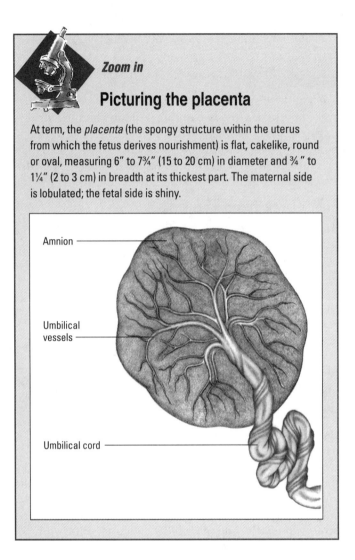

Zoom in

Picturing the placenta

At term, the *placenta* (the spongy structure within the uterus from which the fetus derives nourishment) is flat, cakelike, round or oval, measuring 6" to 7¾" (15 to 20 cm) in diameter and ¾" to 1¼" (2 to 3 cm) in breadth at its thickest part. The maternal side is lobulated; the fetal side is shiny.

Amnion

Umbilical vessels

Umbilical cord

No permanent addition

During early embryonic development, the yolk sac also forms blood cells. Eventually, it undergoes atrophy and disintegrates.

Placenta

The flattened, disk-shaped *placenta*, using the umbilical cord as its conduit, provides nutrients to and removes wastes from the fetus from the 3rd month of pregnancy until birth. The placenta is formed from the chorion, its chorionic villi, and the adjacent decidua basalis.

A fetal lifeline

The umbilical cord contains two arteries and one vein and links the fetus to the placenta. The umbilical arteries, which transport blood from the fetus to the placenta, take a spiral course on the cord, divide on the placental surface, and branch off to the chorionic villi. (See *Picturing the placenta.*)

In a helpful vein

The placenta is a highly vascular organ. Large veins on its surface gather blood returning from the villi and join to form the single umbilical vein, which enters the cord, returning blood to the fetus.

Specialists on the job

The placenta contains two highly specialized circulatory systems:
• The *uteroplacental* circulation carries oxygenated arterial blood from the maternal circulation to the intervillous spaces — large spaces separating chorionic villi in the placenta. Blood enters the intervillous spaces from uterine arteries that penetrate the basal part of the placenta; it leaves the intervillous spaces and flows back into the maternal circulation through veins in the basal part of the placenta near the arteries.
• The *fetoplacental* circulation transports oxygen-depleted blood from the fetus to the chorionic villi by the umbilical arteries and returns oxygenated blood to the fetus through the umbilical vein.

There are two circulation systems in the placenta — one involving maternal blood, and another transporting fetal blood.

Placenta takes charge

For the first 3 months of pregnancy, the corpus luteum is the main source of estrogen and progesterone — hormones required during pregnancy. By the end of the 3rd month, however, the placenta produces most of the hormones; the corpus luteum persists but is no longer needed to maintain the pregnancy.

Hormones on the rise

The levels of estrogen and progesterone, two steroid hormones, increase progressively throughout pregnancy. Estrogen stimulates uterine development to provide a suitable environment for the fetus.

Progesterone, synthesized by the placenta from maternal cholesterol, reduces uterine muscle irritability and prevents spontaneous abortion of the fetus.

Keep those acids coming

The placenta also produces human placental lactogen (HPL), which resembles growth hormone. HPL stimulates maternal protein and fat metabolism to ensure a sufficient supply of amino acids and fatty acids for the mother and fetus. HPL also stimulates breast growth in preparation for lactation. Throughout pregnancy, HPL levels rise progressively.

> The contractions of labor are involuntary at first but are supplemented by voluntary efforts as birth approaches.

Labor and the postpartum period

Childbirth, or delivery of the fetus, is achieved through labor — the process in which uterine contractions expel the fetus from the uterus. When labor begins, these contractions become strong and regular. Eventually, voluntary bearing-down efforts supplement the contractions, resulting in delivery of the fetus and placenta.

When that occurs, the presentation of the fetus may take one of a variety of forms. (See *Comparing fetal presentations.*)

Onset of labor

The onset of labor results from several factors:
- The number of *oxytocin* (a pituitary hormone that stimulates uterine contractions) receptors on uterine muscle fibers increase progressively during pregnancy, peaking just before labor onset. This makes the uterus more sensitive to the effects of oxytocin.
- Stretching of the uterus over the course of the pregnancy initiates nerve impulses that stimulate oxytocin secretion from the posterior pituitary lobe.

Initiating start sequence

Near term, the fetal pituitary gland secretes more *adrenocorticotropic* hormone, which causes the fetal adrenal glands to secrete more *cortisol*. The cortisol diffuses into

Now I get it!

Comparing fetal presentations

Fetal presentations may be broadly classified as cephalic, breech, shoulder, or compound.

Head down

In the *cephalic*, or head-down presentation, the position of the fetus may be classified further by the presenting skull landmark, such as vertex (the topmost part of the head), brow, sinciput (the forward, upper part of the skull), or face.

Head up

In the *breech,* or head-up, presentation, the position of the fetus may be further classified as frank (hip flexed, knees straight), complete (knees and hips flexed), footling (knees and hips of one or both legs extended), kneeling (knees flexed and hips extended), or incomplete (one or both hips extended and one or both feet or knees lying below the breech).

Shoulder first

Although a fetus may adopt one of several *shoulder* presentations, examination can't differentiate among them. Thus, all transverse lies are called shoulder presentations.

Two at once

In the *compound* presentation, an extremity prolapses alongside the major presenting part so that two presenting parts appear at the pelvis at the same time.

Cephalic

Complete breech

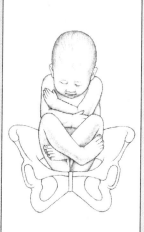

Shoulder

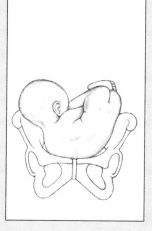

Compound

the maternal circulation through the placenta, heightens oxytocin and estrogen secretion, and reduces progesterone secretion. These changes intensify uterine muscle irritability and make the uterus more sensitive to oxytocin stimulation.

Prostaglandins play their part

Declining progesterone levels convert *esterified arachidonic acid* into a nonesterified form. The nonesterified arachidonic acid undergoes biosynthesis to form prostaglandins, which, in turn, diffuse into the *uterine myometrium*, thereby inducing uterine contractions.

Oxytocin secretions may also stimulate *prostaglandin* formation by the decidua.

Going full circle

As the cervix dilates, nerve impulses are transmitted to the central nervous system, causing an increase in oxytocin secretion from the pituitary gland. Acting as a positive feedback mechanism, increased oxytocin secretion stimulates more uterine contractions, which further dilate the cervix and lead the pituitary to secrete more oxytocin.

Stages of labor

Childbirth can be divided into three stages. The duration of each stage varies with the size of the uterus, the woman's age, and the number of previous pregnancies.

Ticket to drama

The first stage of labor, in which the fetus begins its descent, is marked by cervical *effacement* (thinning) and *dilation*. Before labor begins, the cervix isn't dilated; by the end of the first stage, it has dilated fully. The first stage of labor can last from 6 to 24 hours in primiparous women but is often significantly shorter for multiparous women. (See *Cervical effacement and dilation.*)

The stage is set

The second stage of labor begins with full cervical dilation and ends with delivery of the fetus. During this stage, the amniotic sac ruptures as the uterine contractions increase in frequency and intensity. As the flexed head of the fetus enters the pelvis, the mother's pelvic muscles

Now I get it!

Cervical effacement and dilation

Cervical effacement and dilation are both significant aspects of the first stage of labor.

Thinning walls
Cervical effacement is the progressive shortening of the vaginal portion of the cervix and thinning of its walls during labor as it's stretched by the fetus. Effacement is described in the form of a percentage, ranging from 0% (noneffaced and thick) to 100% (fully effaced and paper-thin).

Bigger exit
Cervical dilation refers to progressive enlargement of the *cervical os* to allow the fetus to pass from the uterus into the vagina. Dilation ranges from less than 1 cm to about 10 cm (full dilation).

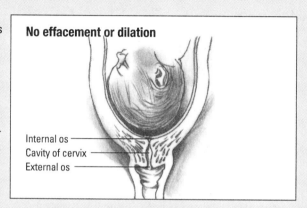

No effacement or dilation

Internal os
Cavity of cervix
External os

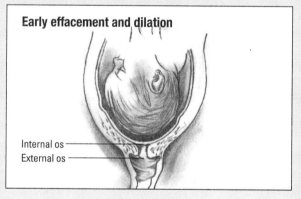

Early effacement and dilation

Internal os
External os

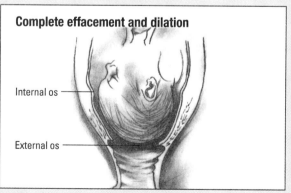

Complete effacement and dilation

Internal os

External os

The first stage of labor can last from 6 to 24 hours in primiparous women, while the second stage lasts about 45 minutes.

Both stages are typically much shorter for women who have delivered before.

force the head to rotate anteriorly and cause the back of the head to move under the symphysis pubis.

The curtain rises

As the uterus contracts, the flexed head of the fetus is forced deeper into the pelvis; resistance of the pelvic floor gradually forces the head to extend. As the head presses against the pelvic floor, vulvar tissues stretch and the anus dilates. At this point, the vulvovaginal orifice may be enlarged surgically by an episiotomy, a small incision of the perineum used to avoid spontaneous laceration.

The star appears

The head of the fetus now rotates back to its former position after passing through the vulvovaginal orifice. Usually, head rotation is lateral (external) as the anterior shoulder rotates forward to pass under the pubic arch. Delivery of the shoulders and the rest of the fetus follows. The second stage of labor averages about 45 minutes in primiparous women; it may be much shorter in multiparous women.

Curtain calls and applause!

The third stage of labor starts immediately after childbirth and ends with placenta expulsion. After the neonate is delivered, the uterus continues to contract intermittently and grows smaller. The area of placental attachment also decreases. The placenta, which can't decrease in size, separates from the uterus, and blood seeps into the area of placental separation. The third stage of labor averages about 10 minutes in both primiparous and multi-parous women.

Postpartum period

After childbirth, the reproductive tract takes about 6 weeks to revert to its former condition in a process called involution. The uterus quickly grows smaller, with most of its involution taking place during the first 2 weeks after delivery.

From red to gray

Postpartum vaginal discharge (lochia) persists for several weeks after childbirth.
• Lochia rubra, a bloody discharge, appears 1 to 4 days postpartum.
• Lochia serosa, a pinkish brown, serous discharge, occurs from 5 to 7 days postpartum.
• Lochia alba, a grayish white or colorless discharge, appears from 1 to 3 weeks postpartum.

Lactation

Lactation (milk synthesis and secretion by the breasts) is governed by interactions involving four hormones:
• estrogen and progesterone, produced by the ovaries and placenta
• prolactin and oxytocin, produced by the pituitary under hypothalamic control.

Hormonal initiation of lactation

As gestation progresses, the production of estrogen and progesterone by the placenta increases, causing glandular and ductal tissue in the breasts to proliferate. After breast stimulation by estrogen and progesterone, prolactin causes milk secretion.

Priming the pump

Oxytocin from the posterior pituitary lobe causes contraction of specialized cells in the breast, producing a squeezing effect that forces milk down the ducts. Breastfeeding, in turn, stimulates prolactin secretion, resulting in a high prolactin level that induces changes in the menstrual cycle.

Opening the throttle

Progesterone and estrogen levels fall after delivery. With estrogen and progesterone no longer inhibiting prolactin's effects on milk production, the mammary glands start to secrete milk.

Pressing on the gas pedal

Nipple stimulation during breast-feeding results in transmission of sensory impulses from the nipples to the hypothalamus. If the nipples aren't stimulated by breast-feeding, prolactin secretion declines after delivery. Milk secretion continues as long as breast-feeding regularly stimulates the nipples. If breast-feeding stops, the stimulus for prolactin release is eliminated and milk production ceases.

Regular breast-feeding provides the stimulus necessary for continued milk production.

Now I get it!

Breast milk composition

The composition of breast milk undergoes various changes during the process of lactation.

First tastes
Initial feedings provide a thin, serous fluid called colostrum. Unlike mature breast milk, which has a bluish tinge, colostrum is yellow. Colostrum contains high concentrations of protein, fat-soluble vitamins, minerals, and immunoglobulins, which function as antibodies. Its laxative effect promotes early passage of meconium, the greenish black material that collects in the fetal intestines and forms the newborn's first stool. The breasts may contain colostrum for up to 96 hours after delivery.

From foremilk to hindmilk
Breast milk composition continues to change over the course of a feeding. The foremilk — the thin, watery milk secreted when a feeding begins — is low in calories but abounds in water-soluble vitamins. It accounts for about 60% of the total volume of a feeding. Next, hindmilk is released. The hindmilk, available 10 to 15 minutes after a breast-feeding session begins, has the highest concentration of calories; this helps to satisfy the newborn's hunger between feedings.

Breast-feeding and the menstrual cycle

During the postpartum period, the woman's high pro-lactin level inhibits FSH and LH release. If she doesn't breast-feed, her prolactin output soon drops, ending inhibition of FSH and LH production by the pituitary. Subsequently, cyclic release of FSH and LH occurs. (See *Breast milk composition.*)

A hold on ovulation

In a breast-feeding woman, the menstrual cycle doesn't resume because prolactin inhibits the cyclic release of FSH and LH necessary for ovulation. This explains why a breast-feeding woman usually doesn't become pregnant.

The cycle begins again

Prolactin release in response to breast-feeding gradually declines, as does the inhibitory effect of prolactin on FSH and LH release. Consequently, ovulation and the menstrual cycle may resume. Pregnancy may occur after this, even if the woman continues to breast-feed.

Quick quiz

1. Each of the three germ layers (ectoderm, mesoderm, and endoderm) forms specific tissues and organs in the developing:

    A.   zygote.
    B.   ovum.
    C.   embryo.

Answer: C. Each of the three germ layers (ectoderm, mesoderm, and endoderm) forms specific tissues and organs in the developing embryo.

2. The structure that guards the fetus is the:

    A.   decidua.
    B.   amniotic sac.
    C.   corpus luteum.

Answer: B. The structure that guards the fetus by producing a buoyant, temperature-controlled environment is the amniotic sac.

3. The progressive enlargement of the cervical os during labor is called:

    A.   dilation.
    B.   effacement.
    C.   lactation.

Answer: A. The progressive enlargement of the cervical os during labor is called dilation.

4. The initial breast milk that is yellow in color is called:

    A.   foremilk.
    B.   hindmilk.
    C.   colostrum.

Answer: C. The initial breast milk that is yellow in color is called colostrum.

Scoring

☆☆☆   If you answered all four questions correctly, fabulous! You're first-rate with the physiology of fertilization.

☆☆   If you answered two or three questions correctly, excellent! You're looking good in the areas of labor and lactation.

☆   If you answered fewer than two questions correctly, don't be alarmed! There is lots of material and plenty of time for a quick review.

Index

A

Abdominal region, 5-6, 5i
Abdominopelvic cavity, 4
Abduction, 55, 55i
ABO blood group, 141, 142
Absorption, intestinal, 218-219, 229-231
Achilles reflex, 82, 82i, 83
Acid-base balance, 177-178, 263-273
   buffers and, 264-269
   disturbances of, 177-178, 265-268
   kidneys and, 270-273
   lungs and, 269-270
   pH and, 263-264, 271i
Acidosis
   metabolic, 268
   respiratory, 266
Acids, 35, 263
Acinus, 165-166, 165i
Actin, 52, 53
Active transport, 19, 259
Adduction, 54, 55, 55i
Adenine, 10
Adenosine diphosphate, 16
Adenosine triphosphate, 16, 259
Adipose tissue, 23
Adrenal glands, 102, 105
Afferent pathway, 78-80, 81
Afterload, 121, 123
Agranulocytes, 136, 137, 137i
Airflow patterns, 173-175, 174i
Airways, 164-166, 165i
Aldosterone, 251i, 252-253, 262
Alimentary canal, 181-186, 182, 184
Alkalosis
   metabolic, 267
   respiratory, 265
Alleles, 12
Alpha cells, 211
Alveoli, 165, 166, 195i
Amines, 108

Amino acids, 224
   and protein metabolism, 235
   synthesis of, 235-236
Ammonia, formation of, 272-273
Amniotic fluid, 303-305
Amniotic sac, 303-305
Amphiarthroses, 64
Anatomic terms, 1-6
   for body cavities, 3-4, 3i
   for body regions, 5-6
   directional, 1-2
   for reference planes, 2-3, 2i
Androgens, 281-282
Angiotensin, 251-252
Anions, 35, 261
Anterior chamber, 91, 92
Anterior horns, 77, 79
Anterior pituitary, 102, 103
Antibodies, 153-154
Antidiuretic hormone, 103, 250-251, 250i
Antigens, 148, 152-154
Aorta, 125, 126
Aortic valve, 116, 118
Apocrine glands, 46
Aponeurosis, 53
Appendicular skeleton
   bones of, 56, 57
   muscles of, 51, 54
Aqueous humor, 91, 92
Arachnoid membrane, 84
Arteries, 124, 125
   of brain, 77, 78
   coronary, 127, 128, 193i
Arterioles, 124
Atomic mass number, 31
Atomic weight, 31
Atoms, 28
   structure of, 29-32
   valence of, 32
Atria, 116, 118
Atrioventricular node, 120

i refers to an illustration; t refers to a table.

Atrioventricular valves, 116, 118-119
Auricle, 95, 96i
Autoimmune disorders, 158
Autonomic nervous system, 85-88
Axial skeleton
    bones of, 56, 57
    muscles of, 51, 54
Axons, 24, 69, 70, 70i

B

Ball-and-socket joints, 66
Barr body, 12
Bartholin's glands, 283-284
Bases, 35, 263
Basophils, 135, 137, 137i
Beta cells, 211
Bicarbonate
    and acid-base balance, 265-273
    formation of, 273
    regulation of, 261, 263
    renal reabsorption of, 272
Biceps reflex, 81, 82, 82i
Bicuspid valve, 116, 119
Bile, 208, 209
Bladder, 246-247
Blastocyst, 298
Blood, formation of, 131, 132-133i
Blood clots, 138-140
Blood components, 134-138
Blood groups, 141-142, 142t
Blood vessels, 124, 125i
B lymphocytes, 136, 137i, 146-147
Body cavities, 3-4, 3i
Body regions, 5-6
Bonds, chemical, 32, 33i
Bone, 56-62, 57i
    blood supply of, 59
    classification of, 56-59
    formation of, 59-62
    functions of, 59
    growth and remodeling of, 60-62, 60-61i
    muscle attachment to, 53
    structure, 58i
    types of, 58
Bone marrow, 147

Brachioradialis reflex, 82, 82i, 83
Brain, 73-77
    arteries of, 77, 78i
    lobes of, 187i
    protective structures in, 84
    structures of, 73-76, 75i
Brain stem, 74-76
Breast, 288, 289i
Breast-feeding, 313-315
Bronchi, 163, 164
Bronchioles, 165-166, 165i
Brush border, 21
Buffer systems, 264-269
Bundle of His, 120
Bursae, 66

C

Calcitonin, 104
Calcium balance, 261, 262
Calvaria, 4
Canaliculi, 58
Cancellous bone, 58
Capillaries, 124
Carbohydrates, 35-36, 222-224
    digestion and absorption of, 229
    metabolism of, 231-235
Carbon dioxide, in acid-base balance,
    265-270
Cardiac conduction system, 119-123, 119i
Cardiac contractility, 121, 123i
Cardiac cycle, 121, 122i
Cardiac muscle, 24, 50
Cardiac output, 121
Cardiac valves, 116, 118-119
Cardiovascular system, 115-128
    structures of, 115-116
Cartilage, 62-63
    ossification of, 60
Cartilaginous joints, 64
Cavities, 3-4
Cecum, 203
Cell
    energy generation in, 16
    structure of, 6-10, 7i
Cell-mediated immunity, 156-157

i refers to an illustration; t refers to a table.

Cell membrane, 9
Cellular metabolism, 16
Cellular transport, 17-20
Central nervous system
    brain in, 73-77
    spinal cord in, 77-84, 79i
Cerebellum, 74, 75
Cerebral cortex, 73, 75
    motor, 80
    sensory, 78-80
Cerebral hemispheres, 73, 75
Cerebrum, 73-74, 75
Cervix, 285, 286
    effacement and dilation of, 310-312, 311i
Chemical bonds, 32, 33i
Chemical compounds, 28, 33-37
Chemical reactions, 32-33, 34
Chemistry, principles of, 27-29
Childbirth, 308-313
Chloride balance, 261, 263
Cholecystokinin, 210
Choroid, 91i, 93
Chromosomes, 9-10, 11-12
    diploid number of, 14
    haploid number of, 14
    and meiosis, 14-16
    and mitosis, 14
Chylomicrons, 231
Cilia, 21
Ciliary body, 91i, 92-93
Circulation, 124-128
    cerebral, 77, 78
    coronary, 127, 128
    pulmonary, 124-126, 172i
    systemic, 125, 126
Circumduction, 54, 55, 55i
Clitoris, 283, 284
Clotting, 138-140
Coagulation factors, 138-140
Cochlea, 96i, 97
Colon, 203
Compact bone, 58
Complement system, 156
Conducting airways, 164-166
Condylar joints, 66

Cones, 94
Conjunctiva, 89
Connective tissue, 23
Contractility, 121, 123
Contraction, muscle, 53-54
Cornea, 91i, 92
Coronary arteries, 193i
Coronary circulation, 127, 128
Corpus callosum, 73
Corpus luteum, 301-303
Covalent bond, 32, 33
Cranial nerves, 85, 86, 86i
Cytosine, 10
Cytoskeleton, 8
Cytosol, 8

D

Deamination, 235
Decidua, 303, 304i
Deep tendon reflexes, 80-83, 82i
Defecation, 218-219
Dendrites, 24, 70, 70i, 71
Deoxyribonucleic acid, 10, 12, 14i
Dermatomes, 40
Dermis, 43-44, 43i
Diaphysis, 62
Diastole, 121
Diencephalon, 76
Digestion, 229-231
    accessory organs of, 207-212
    process of, 212-219
Diploid number, 14
Directional terms, 1-2
Disaccharides, 223
Dorsal cavity, 3-4
Dorsal horns, 77, 79
Duodenum, 185, 186
Dura mater, 84

E

Ear, 95-98
    external, 95, 96i
    hearing pathway and, 98
    inner, 96i, 97-98
    middle, 95-97, 96i

i refers to an illustration; t refers to a table.

Eccrine glands, 46
Efferent pathway, 80, 81
Elastic cartilage, 63
Electrolyte balance, 260-263
Electrolytes, 7, 35
Electrons, 31-32
Electron transport system, 232, 233-234
Elements, 29
    in human body, 30i
Endocardium, 117
Endocrine system, 101-113, 197i
    glands of, 21, 101-107, 102i
    hormones of, 107-113
Endocytosis, 20
Endometrium, 290
    in pregnancy, 298
Endomysium, 50
Endoplasmic reticulum, 8
Endothelium, 21
Energy, 28
Eosinophils, 135, 137, 137i
Epicardium, 117
Epidermal appendages, 44-46
    hair, 44
    nails, 44-45, 45i
    sebaceous glands, 43i, 45-46
    sweat glands, 42, 43i, 46
Epidermis, 42-43, 43i
Epididymis, 276, 278
Epimysium, 52, 53
Epiphyseal plate, 60-62
Epiphysis, 62
Epithelium, 20-21, 22i
Erythropoietin, 253
Esophagus, 183
Eustachian tube, 95, 96i
Eversion, 55, 55i
Exocrine glands, 21
Expiration, 170, 171
Extension, 55
External ear, 95, 96i
External rotation, 55
Extracellular fluid, 257
    electrolytes in, 261
Extraocular muscles, 88

Extrapyramidal system, 80
Eye, 88-95
    extraocular structures of, 88-90
    intraocular structures of, 90-94, 91i
    vision pathway and, 94
Eyelids, 89

F
Fallopian tubes, 285, 287
Fasciculus, 53
Fats, 225
Fertilization, 295-297, 296i
Fetal development, 299-301, 302i
Fetal membranes, 303-305, 304i
Fetal presentation, 309i
Fibrous cartilage, 63
Fibrous joints, 64
Filtration, 20
Flexion, 54, 55, 55i
Fluid
    electrolytes in, 260-263, 261t
    movement of, 257-260
    and solutions, 258
    types of, 256
Fluid balance, 255-260, 256i
    hormones and, 250-253
Food pyramid, 222i
Fovea centralis, 91i, 94
Frontal lobe, 74, 75

G
Gallbladder, 209, 211i
Ganglia, 79-80, 87
Gas exchange, 175-177, 178i
Gastric inhibitory peptides, 210
Gastric secretion, 214
Gastrin, 210, 213-215
Genes, 12, 13
GI hormones, 210t, 213-215
GI system, 181-219, 182i, 198i
    components of, 181-212
    innervation of, 206
    overview of, 181
GI wall, 204-206, 205i
Glands, 21, 101-107

i refers to an illustration; t refers to a table.

Glandular epithelium, 21
Glial cells, 24, 71
Gliding joints, 65
Glucagon, 106, 211
Gluconeogenesis, 234
Glucose metabolism, 231-235
    hormones in, 238-239
Glycogenolysis, 234
Glycolysis, 231-233
Gonads, 102, 106
Granulocytes, 135, 137, 137i
Guanine, 10

H

Hair, 43i, 44
Haploid number, 14
Haversian canals, 58
Hearing pathway, 98
Heart
    anterior view of, 192i, 193i
    blood vessels of, 127i, 128
    contraction of, 119-123
    coronary arteries of, 193i
    posterior view of, 193i
    structure of, 116-119, 116i
    wall of, 192i
Hematologic system, 131-142
Hematopoiesis, 131, 132-133, 133i
Hinge joints, 65
Homeostasis, 35
Hormones, 107-113
    feedback loop for, 110i
    female sex, 288-292, 303, 307-308
    GI, 210t, 213-215
    and glucose regulation, 238-239
    and labor, 310
    and lactation, 313-314
    male sex, 281-282
    and pregnancy, 303, 307-308
    regulation of, 109-111, 112-113
    release and transport of, 108-111, 112-113
    target cells for, 109, 111i
    trophic, 111
    types of, 108
    urinary system, 250-253

Humoral immunity, 153-155
Hyaline cartilage, 63
Hydrogen bond, 32
Hydrogen ions, pH and, 263-264
Hypersensitivity disorders, 157-158
Hypertonic solution, 258
Hypothalamic-pituitary–target gland axis, 112-113
Hypothalamus, 76, 103
Hypotonic solution, 258

I

Ileum, 186
Immune response, 151-157
    nonspecific, 151-152
    protective surface phenomena in, 151-152
    specific, 151-155
Immune system, 145-158, 146i
    components of, 145-150
    disorders of, 157-158
    functioning of, 151-157
    overview of, 145-146
Immunity, 151
    cell-mediated, 156-157
    humoral, 153-155
Immunodeficiency, 158
Immunoglobulins, 153-154
Inclusions, cellular, 8
Incus, 96i, 97
Inferior vena cava, 118, 128
Inflammatory response, 152, 153
Inner ear, 96i, 97-98
Inorganic compounds, 33-35
Inspiration, 170, 171
Insulin, 106, 211, 238, 239
Integumentary system, 39-46. *See also* Epidermal appendages *and* Skin.
Internal rotation, 55
Interstitial fluid, 256
Intestinal absorption, 218-219, 229-231
Intracellular fluid, 256, 257
    electrolytes in, 261
Intravascular fluid, 256
Inversion, 55, 55i
Ionic bond, 32, 33

i refers to an illustration; t refers to a table.

Ions, 260-261
Iris, 91i, 92
Islets of Langerhans, 106
Isotonic solution, 258
Isotopes, 31

J
Jejunum, 185, 186
Joint, 63-66
    ball-and-socket, 66
    cartilaginous, 64
    condylar, 66
    fibrous, 64
    gliding, 65
    hinge, 65
    pivot, 66
    saddle, 66
    synovial, 64-65

K
Ketone body formation, 237-238
Kidney, 242-246, 243i
    and acid-base balance, 270-273
Krebs cycle, 232, 233

L
Labia majora, 283, 284i
Labia minora, 283, 284i
Labor, 308-313
Lacrimal apparatus, 89-90
Lactation, 313-315
Lacunae, 58
Lamellae, 58
Langerhans' cells, 40
Large intestine, 203
    and digestion, 218-219
Laryngopharynx, 162, 163i
Larynx, 162, 163i
Lens, 91i, 92
Ligaments, 56
Limbic system, 76, 187i
Lipids, 36, 224-225
    digestion and absorption of, 230-231
    metabolism of, 237-238

Liver, 207-209
    and glucose regulation, 234
    lobule of, 208i
Lungs, 163, 166
    and acid-base balance, 269-270, 271
Lymph, 148-150
Lymphatic vessels, 148-150, 149i
Lymph nodes, 148-150, 149i
Lymphocytes, 136, 137, 137i, 146-147
Lymphoid organs and tissues
    accessory, 146, 150
    central, 146-147
    peripheral, 146, 147-150
Lysosomes, 8, 9

M
Macrophages, 136, 152, 155
Macula, 91i, 94
Magnesium balance, 261, 262
Malleus, 96i, 97
Mammary glands, 288
Mass number, 31
Matter, 28
Mediastinum, 115, 167
Medulla oblongata, 76
Meiosis, 14-16, 17
Melanocytes, 40
Melatonin, 107
Membrane, plasma, 9
Menopause, 292
Menstrual cycle, breast-feeding and, 315
Menstruation, 288-292
Messenger ribonucleic acid, 11
Metabolic acidosis, 177-178, 268i
Metabolic alkalosis, 177, 267i
Metabolism, 221
    carbohydrate, 231-235
    hormonal regulation of, 238-239
    lipid, 237-238
    protein, 235-236
Midbrain, 74
Middle ear, 95-97
Minerals, 222, 225-229, 226-228t
Mitosis, 13-14, 15i

i refers to an illustration; t refers to a table.

Mitral valve, 116, 119
Mixed glands, 21
Molecules, 28
Monocytes, 136, 137, 137i
Monosaccharides, 223
Mons pubis, 282-283, 284
Motor cortex, 80
Motor pathway, 80, 81i
Mouth, 183, 184
Movement
    mechanics of, 49, 55i
    types of, 54, 55
Mucosa, GI, 204, 205
Muscle, 50-54
    attachment of, 53
    cardiac, 24, 50
    contraction of, 53-54
    extraocular, 88
    functions of, 50
    growth of, 54
    insertion of, 53-54
    origin of, 53
    skeletal, 50, 51i, 54
    smooth, 24
    striated, 24, 53
    structure of, 50-53, 52i
    visceral, 50
Muscle cells, in glucose regulation, 234-235
Muscle fibers, 50, 52
Muscle tissue, 23-24
Muscular system
    anterior view of, 190i
    posterior view of, 191i
Musculoskeletal system, 49-66, 57i
Myocardium, 116, 117
Myofibrils, 50, 52
Myosin, 52, 53

N
Nails, 44-45, 45i
Nares, 162, 163i
Nasal passages, 162, 163i
Nephron, 244-246, 245i

Nerves
    cranial, 85, 86, 86i
    spinal, 77, 78, 85, 196i
Nervous system, 49-98
Nervous tissue, 24
Neuroglia, 24, 71
Neurons, 24, 69-71, 70i
Neurotransmission, 71, 72i
Neutrons, 31
Neutrophils, 135, 137, 137i
Nose, 98
Nostrils, 162
Nucleic acids, 37
Nucleoli, 9-10
Nucleoplasm, 7-8
Nucleotides, 10
Nucleus, 9-10
Nutrient absorption, 218-219, 229-231
Nutrition, 221-229

O
Occipital lobe, 74, 75
Olfactory receptors, 98
Optic chiasm, 94
Optic disk, 91i, 93
Optic nerve, 91i, 94-95
Optic tracts, 94
Oral cavity, 183, 184i
Organelles, 8
Organic compounds, 35-37
Oropharynx, 162, 163i
Osmosis, 18, 19i, 259, 262
Ossicles, 95, 96-97, 96i
Ossification, 59-62
Ossification center, 60
Osteoblasts, 60, 62
Osteoclasts, 61, 62
Osteogenesis, 59-62
Osteoid, 62
Oval window, 95, 96i
Ovaries, 102, 106, 285, 287-288
Ovulation, 290, 291
Oxygen diffusion, 175-177, 178

i refers to an illustration; t refers to a table.

PQ

Pancreas, 102, 105-106, 210-212, 211i
Pancreatic duct, 212
Parasympathetic nervous system, 85, 87-88
Parathyroid glands, 102, 104-105
Parathyroid hormone, 104-105
Parietal lobe, 74, 75
Patellar reflex, 82, 82i, 83
Penis, 276, 277
Pericardium, 116, 117
Perimysium, 52, 53
Perineum, 285
Periosteum, 53
Peripheral nervous system, 85
Peritoneum, visceral, 206
pH, 263-264
Phagocytosis, 20, 152, 155i
Pharynx, 162, 163i, 183
Phosphate balance, 261, 263
Phosphate salts, formation of, 272-273
Phospholipids, 225
Physiologic cup, 91i
Pia mater, 84
Pineal gland, 102i, 107
Pinna, 95, 96i
Pinocytosis, 20
Pituitary gland, 102-103, 102i
Pituitary-target gland axis, 111-112
Pivot joints, 66
Placenta, 306-308, 306i
Plasma membrane, 9
Platelets, 136-138
Pleura, 163i, 166
Pleural cavity, 163i, 166
Polypeptides, 108
Polysaccharides, 223-224
Pons, 74
Posterior chamber, 91i, 93
Posterior pituitary, 103
Posterior sclera, 91i, 93
Postpartum period, 313
Potassium, in acid-base balance, 273
Potassium balance, 261, 262
Preembryonic development, 298, 299i

Pregnancy, 298-299
   fetal development in, 299-301, 300i, 302
   hormones and, 303, 307-308
   preembryonic development and, 298, 299i
   tests for, 303
   uterine changes and, 301-308
Preload, 121, 123
Primitive reflexes, 80, 83
Pronation, 55, 55i
Prostate, 276, 279
Proteins, 37, 224
   digestion and absorption of, 230
   metabolism of, 235-236
Protons, 29
Protoplasm, 7
Protraction, 55, 55i
Puberty
   in female, 288
   in male, 282
Pulmonary circulation, 124-126, 172
Pulmonary perfusion, 175, 176
Pulmonic valve, 116, 118
Pupil, 91i, 92
Purines, 10
Purkinje fibers, 120
Pyramidal system, 80
Pyrimidines, 10

R

Rectum, 203
Red blood cells, 134
Reference planes, 2-3
Reflexes, 80-83, 82i
Regions, body, 5-6
Renal cortex, 242, 243
Renal medulla, 242, 243
Renal pelvis, 242, 243
Renin-angiotensin system, 251-252
Reproduction, 290, 291, 295-315
Reproductive system
   female, 201i, 282-292, 284-285i
   male, 200i, 275-282, 276i
Respiration, 170-177, 171i
   and acid-base balance, 269-270, 271
Respiratory acidosis, 266i

i refers to an illustration; t refers to a table.

Respiratory alkalosis, 265i
Respiratory bronchioles, 165-166, 165i, 195i
Respiratory system, 161-178, 163i, 194i, 195i
   lower, 163-165, 163i
   upper, 161-162, 163i
Respiratory unit, 195i
Reticular activating system, 76
Retina, 91i, 93-94
Retraction, 55, 55i
Rh typing, 141-142
Ribonucleic acid, 10-11, 37
Ribosomal ribonucleic acid, 11
Ribs, 167, 168
Rods, 94
Rotation
   external, 55
   internal, 55
Round window, 96, 96i

S
Saddle joints, 66
Salts, 35
Sarcolemma, 50, 52
Sarcomeres, 53
Sarcoplasm, 50
Sclera, 91i, 92
   posterior, 93
Scrotum, 276, 277-278
Sebaceous glands, 45-46
Secretin, 210
Semen, 279-280
Semicircular canals, 96i, 97
Semilunar valves, 116, 118-119
Seminal vesicles, 276, 279
Sensory cortex, 78-80
Sensory pathway, 78-80, 81i
Sex chromosomes, 11-12
Sex hormones
   female, 288-292
   and labor, 310
   male, 281-282
   and pregnancy, 303, 307-308
Sinoatrial node, 120
Sinuses, 162, 163i

Skeletal muscles, 50, 51i, 54. *See also*
   Muscle.
Skeleton
   anterior view of, 188i
   appendicular, 56, 57
   axial, 56, 57
   posterior view of, 189i
Skene's glands, 283, 284
Skin
   functions of, 39-42
   layers of, 42-44, 43i, 202i
Small intestine, 185, 186, 203
   and digestion, 215-218, 216-217i
Smooth muscle, 24
Sodium balance, 261, 262
Sodium bicarbonate, in acid-base balance,
   264-269
Solutes, movement of, 258-260
Solutions, 258
Somatostatin, 106
Spermatic cord, 277-278
Spermatogenesis, 280
Spinal cord, 77-84, 79i
Spinal nerves, 77, 78, 85, 196i
Spinal tracts, 77, 79
Spleen, 150
Stapes, 95, 97
Steroids, 108, 225
Stomach, in digestion, 183-186, 185i,
   212-215, 214i
Striated border, 21
Striated muscle, 24, 53
Stroke volume, 121
Subarachnoid space, 84
Subdural space, 84
Submucosa, GI, 204, 205
Superficial reflexes, 80, 83
Superior vena cava, 118, 127, 128
Supination, 55, 55i
Swallowing, 212, 213i
Sweat glands, 42, 46
Sympathetic nervous system, 85, 87
Symphyses, 64
Synchondroses, 64
Synovial joints, 64

i refers to an illustration; t refers to a table.

Systemic circulation, 125, 126
Systole, 121

T

Target cells, 109, 111, 111i
Taste buds, 98
T cells, 106
Tears, 89-90, 90i
Temporal lobe, 74, 75
Tendons, 56
Testes, 107, 276, 278
Testosterone, 281-282
Thalamus, 76, 79-80
Thermoregulation, skin in, 40-41
Thirst, 260
Thoracic cage, 167-169
Thoracic cavity, 4, 167-169, 168-169i
Thymine, 10
Thymus, 102i, 105-106, 146, 147
Thyroid gland, 102i, 104
Thyroid hormones, 104
Thyroxine, 104
Tissue, 20-24
   adipose, 23
   connective, 23
   epithelial, 20-21, 22, 22i
   muscle, 23-24
   nervous, 24
T lymphocytes, 136, 137i, 146-147
Trabeculae, 58, 58i
Trachea, 163i, 164
Transamination, 235
Transfer ribonucleic acid, 11
Triceps reflex, 82, 83
Tricuspid valve, 116, 118-119
Triiodothyronine, 104
Trophic hormones, 111
Tunica muscularis, 204-205, 205
Tympanic membrane, 95, 96, 96i

U

Ureter, 246
Urethra, 246-247
   female, 284
   male, 277, 279

Urinary system, 199i, 241-253
   components of, 241-247
Urination, 243, 247
Urine formation, 243, 247-249, 248i
Urine output, 249, 260
Uterus, 285, 287
   in pregnancy, 301-308

V

Vagina, 285, 286
Valence, 32
Vas deferens, 277, 278, 279
Veins, 124, 125
Vena cava, 118, 127, 128
Ventilation, 173-175
Ventilation-perfusion mismatch, 175, 176i
Ventral cavity, 3, 4
Ventral horns, 77, 79
Ventricles, 116, 118
Venules, 124
Vertebral arteries, 77
Vestibule, 96i, 97, 283, 284
Visceral muscle, 50
Visceral peritoneum, 206
Vision pathway, 94
Vitamins, 222, 225-229, 226-228t
Vitreous humor, 91i, 93
Volkmann's canals, 58
Vulva, 282-284

W

Water, body. *See* Fluid.
White blood cells, 134-136

X

X chromosomes, 11-12

Y

Y chromosomes, 11-12
Yolk sac, 305

Z

Zygote, 298

i refers to an illustration; t refers to a table.

Notes

Notes

Notes

Notes